Edge Effects

The

American

Land

and Life

Series

Edited

by Wayne

Franklin

NOTES FROM AN OREGON FOREST

By Chris Anderson

Foreword by Wayne Franklin

University of Iowa Press Ψ Iowa City

EdgeEffects

University of Iowa Press,

Iowa City 52242

Copyright © 1993 by the

University of Iowa Press

All rights reserved

Printed in the United States

of America

Design by Richard Hendel

Photographs courtesy of Oregon

State University College of Forestry

Library of Congress

Cataloging-in-Publication Data

Anderson, Chris, 1955–

    Edge effects: notes from an

Oregon forest / by Chris Anderson;

foreword by Wayne Franklin.

    p.   cm.—(The American

land and life series)

    ISBN 0-87745-419-1,

    ISBN 0-87745-438-8 (paper)

    1. Forest ecology—

Oregon. 2. Hybrid zones—

Oregon. 3. Clearcutting—

Environmental aspects—Oregon.

I. Title. II. Series.

QH105.O7A48   1993

574.5'2642'09795—dc20

                                93-20638

                                CIP

97 96 95 94 93 C 5 4 3 2 1

97 96 95 94 93 P 5 4 3 2 1

FOR JOHN, MAGGIE, AND TIM

Contents

Foreword *Wayne Franklin*

Buying his dream house on the fringe of Corvallis, Oregon, several years ago, essayist Chris Anderson hoped to find in the forest there what eighteenth-century urban writers called "the joys of rural retirement." He hoped to escape the gnawing cares of the city, of which even Corvallis may give a taste, and cultivate gentler and more reflective moods. All too soon, however, the realities of his family's new situation sank in, and Anderson found how relentlessly care may pursue the exurbanite. Repair by repair, hefty check by hefty check, he discovered, as "Mr. Blandings" did before him, that dreams can easily sour. Like Mr. Blandings, too, he quickly became the hostage of contractors, of all fates perhaps the most modern and most quintessentially urban. Rarely is one more aware of abject dependency on the costly skills of others.

There were compensations. As the repair work expanded in scope and cost, Anderson at least could look outside the windows toward the always fresh forest. He was embowered, an eighteenth-century poet might have said, by the seemingly endless trees:

> Below me trees unnumbered rise
> Beautiful in various dyes.

So wrote John Dyer in "Grongar Hill." He went on, in his world-weary way, to praise the general view from that eminence:

> Ever charming, ever new,
> When will the landscape tire the view?
> The fountain's fall, the river's flow,
> The woody valleys warm and low;
> The windy summit, wild and high,
> Roughly rushing on the sky.

There on his hillside in Oregon, the firs rising ever-green outside his windows, Anderson might have sung that song, too. But not for long. The McDonald Forest is an institutional tract as much as an ecological one, belonging to the same university for which Anderson works. Designated as a "research forest," it was about to be the site of a big, seemingly sinister research project. The contractors hardly had finished their work.

Little did Anderson then know that his family was part of the experiment. In addition to studying the ecological effects of clear-cutting patches of timber, the researchers wanted to see how the urban fringe dwellers might be affected, too. The irony of escaping the city only to find oneself in an open-air laboratory was not lost on Anderson or his neighbors. They put aside the delights of retirement long enough to organize and to vocally oppose the disruption of "their" forest. Public meetings became heated as the foresters and the forest-dwellers doubted each other's motives and sanity.

After the clear-cutting occurred, however, Anderson began to see beyond the mobilized politics of that confrontation. He has the honesty to tell his readers in this book how little the timber cutting really did affect his life. He had shared in the original alarm of his neighbors, shared with them the sense that deep assumptions they all clung to were being violated. Then, as the sound of the chain saws became familiar, as he came to know the individuals on the "other" side, as he began to walk through the sunlit patches of open ground that the experiment created, he saw beyond simple oppositions altogether. He had to admit, in fact, that he positively liked some of the changes, and that he barely noticed the great bulk of them anyway.

More important, he came to a new understanding of the McDonald Forest, forests in general, the larger landscape, and the role that we all play in shaping what we naively refer to as "nature." The forest behind his house was, in fact, an artifact rather than a natural domain, the result of "fire suppression" that took place when settlers ended the Native American practice of wide-scale field-burning and the fir trees reclaimed the prairie. This forest had been constructed, as surely as his house on its edge, and the more he learned of forests and the environment the more he came to see McDonald as emblematic of the jumbled complexity of our presence on earth. If the fir trees he so deeply enjoyed had resulted from human intrusion, then how could this new intrusion be attacked solely *as* an intrusion? Wasn't his house intruding? Wasn't he when he walked through the woods?

The old binary oppositions crashed to the earth as he learned more about the thick layering of the world, and a creative second growth of thought shot up in their place. As Bill McKibben and Carolyn Merchant and the practitioners of the new eco-history push us into what we might aptly term the "post-natural condition," we may lament the loss of an older, seemingly more innocent connection to the world. What they argue also holds enormous promise, however, for giving us a better sense of ourselves, our history, and our planet. To the extent that we can understand nature as both a mental construct and a material construction, we may in fact overcome some vexingly old problems. Seeing the world this new way allows us to trace out our responsibility for its condition. If we can move beyond our old mental and material assumptions, then we may be able to alter our thinking and our behavior for the better. It may be too late, McKibben suggests, for us to restore control of some phenomena such as the weather to the self-regulating forces that once dominated them. There's no "outside" anymore, and that's that. Indeed, in many ways there never was, except insofar as we convinced ourselves that the human world had an edge, and that beyond that edge lay what it was convenient, or useful, or exciting, to call nature. Because we attributed to nature the ability to heal itself, and spoke of our own puniness in comparison to nature's great energies, we allowed ourselves to do whatever we felt like doing. Knowing how our structures of thought interpenetrated with our economic behavior and interests, especially during the Industrial Age, will help us to understand how in fact we always affect the world. If we must cling to the concept of nature, it now must be ample enough to include ourselves inside it.

But the social benefits of a "post-natural" view of the world are potentially even more important than the environmental ones. If nothing exists outside the effect of human art, then we cannot escape citizenship in the world community. Those who have fled the city to escape the city's ills will find, if they are as unblinking as Chris Anderson, that they have brought the city with them—and, indeed, that the city has inevitably preceded them on the very ground of their escape. Nature as a concept and a mythic place has long been used to validate a certain anti-social element in American life. We have tended not to take seriously the inevitable frictions and abrasions of our existence together, here on this earth, as long as we thought we individually could evade them by running off to the woods. By demanding that nature give us personal solace for the wrongs of the social world, however, we have made ourselves part of the problem. By dint of the material demands we place on the urban fringe, to say nothing of our emotional ones, and by dint of our flight from the ills whose solution is our responsibility, too, we in fact have worsened both the environment and society. The university foresters' experiment entered Anderson's field of vision at just the right moment. It came as a reminder, so to speak, of the inescapable sociality of our post-natural life. Political organization of the sort he chronicles is another such reminder. So, more intriguing, is his discovery that the ills everybody in his neighborhood feared did not always materialize. They foresaw what their cultural scripts triggered in their imaginations, figuring events as an invasion of jealously guarded personal rights by some representative of a society beyond whose clutches they had thought to transport themselves. The enemy, of course, was also themselves, not least because of this very habit of mind.

Anderson eventually understood all these questions, but finally he transcended them. A different drama began to play itself out as he settled in along the border of the McDonald Forest and even moved, occasionally, beyond it. The edge he came to inhabit was not the literal boundary between Corvallis and the forest, where he thought at first that he was relocating, but rather the more elusive line between assumption and perception, engagement and inwardness, that runs, as it were, through the psyche. An essayist by affirmation rather than default, Anderson values the essay for its ability to render in language the dispersed, often nonlinear, noncumulative insights of individual life. Like an ancient gatherer exploiting the riches of a forest edge, Anderson wanders the fringes of his late-twentieth-century life, picking here and there the fruits that shine forth with color in the dusk. In ways that deepen and enrich as one inhabits—rather than just skims through—these essays, he shows an uncanny ability to take ecological ideas as both truths and metaphors for the wonderfully fertile spaces of the essayist's patient mind. There is, in fact, a complex ecology within the structure not only of individual pieces here, but also of the book as a whole. Like the forest, it rises as an artifact of diversity, with its separate stands of peculiar growth, its overall symbiosis, its passage of energy through complex cycles. In its artistry, its language, its contemplative honesty, *Edge Effects* mimics the literal sites of its own birth. In its gathering of themes, its layering of moods, its alternate illuminations and dusks it is of the world as well as about it.

Edge Effects provides, too, a glimpse of what the new nature writing of the post-natural era may become. In its classic nineteenth-century form, nature writing built on the false romantic distinction of culture and nature, a distinction that for much of our own century we have taken, reactively, as a kind of law. At a time when the urbanizing, industrializing economy of which we are all the inheritors fiercely aggregated and regimented our ancestors, books of nature writing served as an outlet for the nostalgia people recently removed from the land felt for their old domains. This relatively new literary form, in other words, served relatively old social purposes. At the same time, nature writing promoted the emergent view that personal access to nature—in the city park, the suburban yard, the garden, along the roads to which autos soon would deliver city dwellers—"solved" the problem caused by urban concentration. By submerging social issues except as a point of departure or the occasional target of a nature-lover's diatribe, the other nature writing pretended that solutions to the problems of its age lay within the reach of solitary individuals. It was, in fact, profoundly attuned to the individualistic and acquisitive ethos of industrial society itself. In the form of the book, at a minimum, one could purchase from the market the simulacrum of the world everyone had lost.

Chris Anderson refuses, here, to offer the latest version of that old panacea. His honesty includes the fundamental admission that nature is no more unified as a subject than is the writer who may, at times, look in its direction. Having engaged the McDonald Forest so vigorously at the outset, Anderson slowly disengages from it. He comes to see that if he is himself a forest, the forest is as much a social product as he is. Keeping us on the perpetual edge of such interlinked domains, he will not let us rest with the purity of the old "Nature"—or, for that matter, the purity of the old "literary" language. If his prose is redolent of the woods, his woods is itself a "forest of voices." Seeking refuge in either part of his complex equation, we find ourselves reminded of the other. He has the grace to help us escape by insisting on bringing us home. By showing us that there is no place to escape to, he liberates us from the prison of our old mythologies. Modestly, with a diffident gesture, he gives us back the earth.

To the Reader

Think, live, be: next try to express scrupulously

what you think, what you are living, what you are.

—Henri de Lubac

The essays that make up this book describe the first few years of my life on the edge of McDonald-Dunn Forest, a 12,000-acre research forest outside Corvallis, Oregon. The house we live in is an ordinary house, pleasant neighbors each acre and a half, Timberhill Mall only minutes away. And yet we live on an edge. A forest stretches out behind us, full and rich and deep, as wild in places as the Cascades or the Coast Range, and living near it these past few years I've felt myself change. I've had adventures, in the woods and in my mind. The pieces I've arranged here form a kind of journal of those adventures, a record of the relationship between the forest and the life I live on its boundary.

By "edge effect" ecologists mean the tendency toward greater variety and density of plant and animal life in the transition zone between two plant communities—a prairie and a forest, a clear-cut and a stand of old-growth. A boundary is a very congenial place for species like gopher and deer. There are always two systems close at hand, ready to exploit and move between, the forest, say, for shelter and the meadow for food all along the large surface area of the boundary itself. For me the "edge effect" has meant a greater variety and density of experience, a multiplying of perspectives. Life is fuller here on the edge, and harder. There's more beauty and more tension, greater solitude and greater obligation.

In a way *Edge Effects* is my version of "The Socio-Economic Impact of Harvesting Techniques on Residents on the Urban

Fringe," an economist's study of the clear-cutting of our hillside. The shock of that unexpected harvest soon after we moved to the forest compelled the first few essays that follow. Like the wildlife biologists and silviculturalists and forest engineers and sociologists studying every aspect of the cutting, I, too, wanted to gather data. But my tool is the personal essay, and the data I've gathered concern my own feelings, impressions, and glimpses of beauty. I've simply tried to be empirical: these are the things I've actually thought and felt during the harvest and since, in more or less chronological order. These are the field notes of a several-year study of my life and mind in the midst of trees.

Only about half of them qualify as "nature writing," I think, at least in the sense of pure description and evocation of a landscape. The other half are autobiographical, simply stories of my life, though the forest is always there as a backdrop, the place I go to walk and think as I try to map out this inner territory. That's the general movement of the book, in fact, away from the forest itself and toward the forests of mind and memory, a movement that reflects how my interests have changed since I first started writing. My plan in the beginning was to do a cultural history or journalistic exposé, something practical and somehow objective, but more and more I found myself drawn to other, more personal concerns, drawn to silence and the interior life, and gradually I began to understand that this was my real theme, this being pulled inward, toward the self, this frequent guilt and uneasiness about being pulled inward, this tension between reflection and responsibility. My question, I realized, is what to *do*, how to *act*. The forest evolved from subject to setting to metaphor, to put this another way, though finally I came to see these perspectives as interdependent, the outer and the inner world continually collapsing into each other, exchanging positions. The lesson I learned is that simplicity is always cultivated, nature always a paradoxical idea.

Eventually I decided to write each essay as a separate piece and as things happened, not worrying at the time about its fit with other pieces, trying to fulfill the demands of that particular moment, that particular idea and form. I allowed myself to re-think, overlap, simply explore what interested me. And though I've since worked hard to sequence and blend the finished pieces into a unified book, even fictionalizing a little now and then to smooth out transitions, I've also hoped to preserve that sense of occasional-ness, of individual responses to individual moments. That's the way life seems to me, broken up into moments of joy or struggle. Or better: life is like a forest, complex and multiple and varied, the smallest part of it important, all of it changing and in process, every layer of it continually dying to newer, higher forms that feed off and elaborate the old. In this sense *Edge Effects* might be seen as a forest of essays, the parts finally absorbed in a larger ecology of idea and image.

Nothing is ever lost—or resolved. The old issues keep return-ing, subtler and deeper. The same patterns keep reasserting them-selves, breaking up and reforming, breaking up and reforming from stage to stage to stage.

What I've come to respect most about the form of the essay is its refusal to tyrannize over the variety of our impressions. The essay is open to the way things really seem. Deeper unities can develop. A kind of freedom is possible. For the essayist meaning can best be found in story, in the ebb and flow and messiness of what happens to each of us day to day, not in theories or systems or theologies. What interests the essayist is not experience itself but what seems to be revealed there, and only there, not the self but the truths that only the self can know. And this is the key. What's required is a celebration of moments, an acceptance of change. It was only when I resigned my larger ambitions to the modesty and precision of the essay that I was able to look around and see where I really was, where I'd been all along, deep in the forest after all, deep in the wilderness of my own life.

My thanks to Frederick Buechner for his example and encouragement; to Patrick Jordan, for publishing my essays in *Commonweal*; to Carl Klaus, for all the long phone conversations; to Peter Elbow and Betsy Wallace, for their sympathy and encouragement on an earlier manuscript; and especially to Lex Runciman, for his friendship and his wisdom about writing.

My thanks to the writing group for their generosity as readers—Marion McNamara, Kathy Moore, Steve Radosevich, Lex Runciman, Bruce Weber, and Gail Wells—to Kathy in particular for her thorough, insightful reading of the whole first draft; to Steve, a professor of Forest Science, for his technical advice on the final draft. (Whatever imprecisions or errors of fact remain are my doing, not Steve's.)

My thanks to the Oregon Council for the Humanities and the Sustainable Forestry New Paradigm Working Group at Oregon State University for funding my forestry research and to everyone who took the time to talk with me or walk with me in the woods: Jim Boyle, George Brown, Mark Brunson, Carol Chambers, Jeff Garver, Phil Hayes, Becky Johnson, Bill McComb, Scott Peets, Dave Perry, John Tappeiner, and especially Bob Zybach.

My thanks to the office staff in the English Department for their expert help printing the manuscript.

My thanks to the following publications for permission to reprint several of these essays (usually in revised form): the *Sunday Oregonian* (part I of "Forest of Voices"), the *Georgia Review* (part II of "Forest of Voices"), *Crosscurrents* (part II of "Excursions"), and *Commonweal* (parts II and III of "Life on the Edge," parts I and IV of "Excursions," and part V of "Deeper in the Forest").

My special thanks to Wayne Franklin, editor of "American Land and Life," for his sympathy, scrupulousness, and acumen as an editor—*Edge Effects* is a much different, and a much better book, because of his skill; and to Paul Zimmer of the University of Iowa Press, both for forcing me to rethink what I was doing and for allowing himself to be convinced about the final arrangement of the manuscript.

My thanks to Andy Dufner, S.J., for direction both spiritual and literary and for the example of the life he leads.

Above all, my thanks to Barb—my thanks, my respect, and my love.

Forest of Voices

One impulse from a vernal wood

May teach you more of man,

Of moral evil and of good,

Than all the sages can.

—Wordsworth

It was a fine October day, the sun was shining, and walking into the open house I had the impression of trees all around, of a forest and of sunlight coming through a forest. What surprises me now is how big a decision we based on my first foolish gladness at the sight of those trees. The house was large and pleasant, with a large lot bordering the university's research forest, five miles out of town. The agent went on and on about its virtues. But I was sold on the spot, unthinkingly. Later I denied acting out of mere longing for a green world, but there it was, in the moment: a certain progression of rooms, a certain quality of light, but most of all the fir seen through the picture windows, the sense of *forest*.

The complications, of course, began right away, first with the house and then in the forest itself, the smaller crises foreshadowing the larger in a way. But my private ecology was also taken up and changed by the larger one, expanded, and that's what this story is finally about. It's about the impossibility of escape, a forced move beyond the self to the obligations of community.

First there were the ritual wee-hour agonies about interest rates and property taxes and the materialistic values we had always resisted, until now. I walked around wide-eyed and pale for weeks, tense with ironies, our projected house payment repeating itself in my mind like a mantra. To get to our green world we had to pass through the manila world, the endless paper world, of real estate contracts and loan applications. Even to consider the simple life in the country we had to think harder about money and status and possessions than we ever had in our little starter house—our little house off the arterial, cheap and tacky enough to free us from worry about the bottom line.

Once we crossed that threshold and made our embarrassing weekend calls for help, once we weathered thirty-six separate showings and six weeks of daily bathroom spiff-ups to finally sell the old house, the inevitable glitches developed, unique only in their particulars.

The people who bought our house were from New Caledonia, a former French colony off the coast of Australia, and after signing what he thought was all the paperwork, the man returned to New Caledonia and his job as a tuna consultant for the Micronesian fishing industry. He was at sea when the bank discovered he'd signed some of the wrong papers, and when he was finally contacted, two weeks later, he sent the new documents back in French, unacceptable to the mortgage company. There were several more weeks of cables and speculations about the location of the southern fleet before the forms were satisfactorily executed, and in the meantime interest rates had risen a full point. We hadn't locked in when we could have, convinced by President Bush's confident speech to the building association that rates would fall—the first and only time we'd listened to him.

Then our bank was seized by the federal government. It was business as usual the next day, just a reorganization of assets in light of some bad loans, everyone's money insured, but first there was the shock of hearing the news on the eleven o'clock report and another sleepless night feeling like we were the ones at sea, tossed by giant waves.

Then the home inspection arrived, a computerized inventory of all the "delayed maintenance" we'd begun to suspect at the new house, pages of particulars. The roof, in fact, was going. The heat exchanger on the furnace was cracked. Four of the joists holding the livingroom over the ravine had rotted almost through. Our agent managed to hold us together as the proposals and counterproposals shuttled back and forth, the various contractors putting in their bids, but I was running on dumb instinct at that point, weakened by adrenaline surges. Nothing made sense.

It was only at the end, when the final papers were signed and the keys in hand, that we found out about the ten Yorkshire ter-riers, and the cats and rabbits and chickens and cockatoos these people had let loose in their house. They'd all been caged at a neighbor's before, the smell disguised by flowers and potpourri.

Seventy-six hours in advance of the movers, dull with panic, we pulled up twenty-five hundred square feet of carpet with our bare hands, a gold shag so foul the carpet installers refused to haul it away, then shellacked every inch of the stained under-flooring with a twenty-five-dollar-a-gallon sealer primer de-signed for disaster restoration. That was when I was swarmed by the damp wood termites, as I pulled up the final corner of carpet in the last downstairs room, their little orange bodies getting trapped in the hair on my arms. I just laughed. It was fine. We called in another contractor, wrote another four-figure check, and went back to the shellacking.

Suing didn't seem worth it. The inspection contract was full of loopholes, and we learned that the sellers had been in a head-on car accident the day after they moved, barely surviving. Besides, the whole fiasco was shaping up as another loss of innocence experience—a case of karma revenge for my initial naivete—and there was odd comfort in surmising that pattern once again.

The lessons deepened once we moved in. In the first heat of desire I hadn't realized that Highway 99 was just down the hill about a mile, hidden by trees, and that the sound of the log trucks and the commuter traffic would rise up around us every day, fainter than in town, but still there, a ceaseless, dispiriting hum. In the first flush of infatuation I hadn't realized how many other houses were in our neighborhood, camouflaged as they are by trees, that we would still live in a neighborhood, with mail-boxes and barking dogs and stereo bass seeping through the summer air. At times it seemed that the only real difference between the old life and the new was that we spent hours in the car, hours planning trips and errands—that moving to the country had only implicated us further in the constant traffic we'd left town to escape. The inevitable drabness of being, Emerson called it, as inevitable on the margin of McDonald Forest as in line at the mall.

There is no escape. That's one home truth I had to relearn, one idea I organized my irritations around. Simply: nothing is perfect. Our first responsibilities are inner—moral and imaginative—wherever we happen to live.

Another realization. Part of the pastoral fantasy triggered by the windows and the trees was some hold-over, pioneer vision of self-sufficiency. But instead of single-handedly butchering deer and lashing lodgepoles and putting in the crops, I spent the first few months of living in the country mutely nodding as shake-roof savers, carpenters, sheetrock plasterers, carpet installers, shower-tile replacers, plumbers, septic tank pumpers, well inspectors—whole committees of people—gravely trooped through my house and explained the extent of whatever damage in respective jargons too arcane to follow.

I am utterly dependent even now that the major repairs are more or less worked out. I have neither the skills nor the desire to make my own clothes or bake my own bread. My garden is a hobby, my acre and a half merely aesthetic. My work is in town, where I commute every day, a specialist like the rest of us. Of course. It's not that I consciously expected any of that to be different. It's that I unconsciously did. It's that deep down, in ways I hadn't recognized before, I was moved in part by some instinctive drive for solitude and wilderness.

The rescue teams came and went as I mulled these things over. By midsummer the fever of getting and spending subsided, our life returned to routine, and I started enjoying the house for what it really was, for its modest scenes and graces. From the window of my study I could see oak, willow, rhododendron, blackberry, apple, fig, birch, fir, and all the busyness of birds, the flickers and nuthatches and chickadees. From the window of the livingroom I could see the maple in the foreground of the larger forest, leafing out in the spring and coloring in the fall, swelling and fading with the seasons. From the deck I could see stars wheeling in a brilliant black sky.

And there was still the forest behind us, the 12,000 acres of the research forest climbing the hills to the west and north—a forest all around. Every day I could go exploring there, learning the names of wildflowers, hiking over the ridge into the Soap Creek Valley. From my own front door I could walk for miles among fir and madrone and circling hawks, never seeing a house or farm. Whatever else happened, I thought, we lived at the edge of a forest, and I could lose myself in it, disappear in it. The summer wore on and I began to feel at home on the trails, at ease. Looking out from the logging roads on the western slope, I recovered some part of that first wild hope among the trees.

It was late August when we learned about the clearcutting. The days had been sunny and warm. Smoke curled from barbeques on every evening deck. But we'd been on the logging road—had talked with the graduate student pulling on his boots on the tailgate of a forestry truck—had heard him say the word, "clearcut," gesturing vaguely. The College of Forestry had been planning the harvest for months but had held off notifying the neighbors until days before the cutting was to begin. We were part of the experiment—the sociological effects of harvesting techniques on residents on the urban fringe—and earlier notification would have biased the results.

The project was all laid out for us at a public meeting several days later, the researchers nervously addressing a packed house. It's actually a study of clear-cutting alternatives, they said, involving ten plots running east and west along the boundary between the forest and our neighborhood. Seventy-three acres are for clearcuts, the experimental "control"; 44 for "shelterwood" or "two-story" cuts, a technique that leaves a few existing trees in the harvest zone; and about 109 for "patchcuts," miniclearcuts carved into the interior of the larger forested area. The effects on birds and mammals are another focus of study, as well as operational logistics and the possibilities of reforestation. So. Logging was to begin in two days, continuing through February, trails closed until further notice.

We argued and questioned, asked to see the computer-enhanced photographs again, but we knew we had little recourse, legal or otherwise.

In a way it was all of a piece, easy to fit into the plot I had constructed to explain my recent experience: mistaken assumptions, panels of experts, what to make of a diminished thing. But the dread felt larger. More was at stake now than my private aesthetics, and the damage wasn't as easy to repair as simply reinforcing joists or replacing carpet. The trees to be cut were seventy to one hundred years old, irreplaceable in our lifetime however successful the reforestation. More was at stake than my personal maturity. The national crisis had come to my backyard, literally, my private picaresque absorbed into the larger drama of things.

I vacillated at first between resignation and forced faith in the powers that be; tried turning all my energy back into the garden and the yard. But I could hear the logging whistle every morning as I got out of bed, could hear the trees crack and groan as they fell, could smell the pitch of the cut stumps hundreds of yards away. Weekends I hiked the trails monitoring the loggers' progress, scrambling over trees I used to walk in the shadows of. Gradually I started attending meetings, writing letters, doing research.

My response to the loss of those trees was just as visceral and instinctive as my first gladness at the sight of them, and it moved me into action once again. For the second time in a year I was forced out of myself and onto a still wider sea. For the second time my desire for solitude had led me into community—the community of scientists and forest managers and neighbors I found myself nodding and listening to this time, trying to understand.

There are no easy answers here, no villains. The research itself seems interesting and important, though I'm convinced now that in part it was an excuse for making the two and a half million dollars the College of Forestry netted from the harvest. Except for the clearcut, I have to admit that the effects of the harvest, though dramatic enough, are not as catastrophic as I first feared. The patchcut forest is simply more open than before, full of vistas, but this is only the first stage in a thirty-year experiment. The chain saws may be back in a few years to take another percentage of the volume, and they could be back sooner, depending on the politics in the College of Forestry and the demand for revenue.

All I know is that whatever the future, the forest, like the house, is no longer a refuge but a responsibility. More than the house, ironically. I have much less control over what happens in the woods than what happens in my daylight basement, but it's the woods I think I can't trust to the experts or give up as another example of life's imperfections. The problems are too big to be philosophical about or to hide from in ineptness. It's time to learn. It's time to act.

In *The End of Nature* Bill McKibben argues that the greenhouse effect has killed nature as idea and fact—nature in the sense of an otherness, something not human and created. Pumping carbon dioxide into the air has forever changed the atmosphere, raising the temperature three to eight degrees by the next generation, inexorably, everywhere, so that no breeze will ever blow again that we can take the same solace in:

Now that we have changed the most basic forces around us, the noise of the chain saw will always be in the woods.

We have changed the atmosphere, and that will change the weather. The temperature and rainfall are no longer to be entirely the work of some separate, uncivilizable force, but instead in part a product of our habits, our economies, our ways of life. Even in the most remote wilderness, where the strictest laws forbid the felling of a single tree, the sound of that saw will be clear, and a walk in the woods will be changed—tainted—by its whine. The world outdoors will mean the same thing as the world indoors, the hill the same as the house. . . . Yes, the wind still blows—but no longer from some other sphere, some inhuman place.

Reading McKibben the past few days, reflecting on my experience here the last year, I've started to understand the sound of the chain saws in the forest I live by as symbolic of the problems happening in every forest on the planet. Maybe I didn't make a mistake buying my house, moving to the edge of these woods. Maybe there are no better places to seek refuge than McDonald Forest because there are no better places.

If McKibben is right, I need to change my perspective. Maybe the only way for us to mature as a culture—for us to survive—is not to repress our love of the forest as childish and utopian but to acknowledge it as genuine and then to act on it publicly, not settling for merely private insights.

I think, as Wordsworth said, that our being's heart and home are with infinitude and only there. That's one sad benefit of our worry about the greenhouse effect: that it reminds us of the limitations of the physical and the source of all love and honor. But this isn't an argument for resignation. If McKibben is right there is no escape and we must, therefore, turn and fight—turn and consider. We must make our stand here, in our own backyard, in the backyard of the earth, which is everywhere now, unavoidable, the place we've been given to inhabit, however fouled in the past, however delayed the maintenance, the place we've been given to love and make the best of or there's no place left at all.

The forester had been dozing in the forestry truck, waiting for the satellites to come up. He was parked off the road in the northeast corner of the forest, blocking the gate to a meadow, something like a geiger counter set on a tripod next to the truck. He said there'd be four satellites in another hour—invisible in the afternoon sky—and that by bouncing signals off each he'd be able to establish the corner for a "brass cap survey" for a GIS map of this part of the forest. He showed us the actual brass cap, the size of a cookie, cemented in the ground.

A few minutes later Bob and I walked in the prairie behind him, a remnant of the prairie that covered this land for hundreds of years, most of it taken over now by the forest. It was June, and the wind blew across onion flower, clover, wild sunflower, Hooker's pink. Then: the suggestion of wagon ruts, as if two people had been walking side by side ahead of us—as if the wind were blowing harder in two narrow rows, making faint corridors in the grass. That's the Applegate Trail, Bob said to me, pointing, the southern route of the Oregon Trail. A hundred and forty years ago the horses and oxen and wagons would come up over the hill to the south, he said, curve to the west to avoid the marshy valley floor, then spread out right there, right through where the forestry truck is sitting, on their way north. You can see evidence of the trail in that row of fruit trees and oak over to the right, too, the vegetation patterns uniform and straight seen from the air. Farther down, among ash and cottonwood, we saw even clearer ruts in the widening of a cutbank on Soap Creek.

Waiting for satellites on the Applegate Trail. It's an image that sticks with me not just because of the historical irony, that brief slipping of perspective, but because it suggests how plotted and pieced and inscribed this forest is, how overlaid with the human. Past and present, the forest is everywhere enmeshed in human mapping, human measuring.

The forest in fact is an invader, I learned, the prairie much older and in some sense original. If I had looked up from my wagon 140 years ago I would have seen nothing but waving grass and an occasional isolated oak or fir. The forest I see now is the product of human intervention, existing in this form only because of the ecological impact of the settlers who flowed up this trail and into the valley.

Seen on a map Oregon State University's McDonald-Dunn Research Forest looks like a lopsided wing, the apex pointing east, a ridge line of hills and small peaks defining the V. The hills are the beginning of the Coast Range; the wheat and grass fields of the Willamette Valley flatten out to the east. The lower part of the wing is McDonald Forest, 6,800 acres running southwest to northeast just five miles north of Corvallis. The upper part is Paul Dunn Forest, 4,073 acres running southeast to northwest. It's a mixed Douglas-fir and hardwood forest now, typical of this side of the Coast Range, the fir always in the process of crowding out the oak and maple and madrone. Hazel and oceanspray compete with seedlings in the underbrush. The College of Forestry began acquiring the land in the late twenties with money given by Mary McDonald, the elderly widow of a mining and timber baron in San Francisco, buying up logged-over or tax-delinquent tracts piece by piece for use as a research laboratory. After sixty years of management and experimentation the forest is still a "mosaic," medium-sized clearcuts alternating with 15-, 30-, and 50-year-old plantations up and down the V. A few pieces of 100- to 140-year-old trees remain, and there are small sections of old-growth on the upper northern draws.

I didn't know any of this until recently, even after I bought the new house on the boundary of the forest (at the top of the lower wing, near the apex). I am a person of atmospheres and moods, not by nature interested in science and historical fact, lacking a sense of topography. The forest was just a line of timbered hills I could see from town, a place to hike and brood when life got too complicated. I moved to its boundary for all the sentimental reasons people move to forests, for silence and solitude and simplicity.

But then the College of Forestry decided to log part of the hillside behind the house and I suddenly found myself attending that first meeting of homeowners, listening to scientists and forest managers trying to explain some of the complexities beneath the beautiful surfaces. There must have been over a hundred people packed into Peavy Lodge that evening, all of us in L. L. Bean khaki and flannel, it seemed, and all of us concerned, uneasy. The scientists and managers talked about "gap dynamics" and "biodiversity assessment" and "aesthetic viewsheds," showing us computer-enhanced photographs of the "targeted" timber "units" before and after "treatment." Neighborhood environmentalists stood up angrily, reading from prepared statements arguing for the integrity of the ecosystem. Several other homeowners spoke up for the forestry staff, to scattered applause, repeating the available cliches on jobs and the timber supply.

I had moved to the forest for silence but found myself immersed in words. I had moved for experience but found myself struggling with competing theories. I had moved for solitude but suddenly was trying to situate myself in a tense, divided community. Interviewing forestry faculty, tromping through the poison oak, or driving the logging roads, I learned over the course of one summer that the forest I live near is a forest of voices, of language and ideas.

Anxiety motivated my researches at first, the hope that knowledge would bring perspective at least, if not evidence for arguments and strategies for influencing policy. I joined the Sustainable Forestry New Paradigm Working Group at Oregon State, an interdisciplinary faculty research group close to decision-making in the College of Forestry. I started asking around. From the beginning the forest intrigued me as a subject for writing, too. I knew that writing an essay was probably the best way I had of getting people to pay attention to the problem; and more and more the dilemmas and conflicts of forestry began to interest me in themselves, take on a life of their own. If nothing else I knew that writing would help me handle the crisis personally, help me order my feelings. In a sense, I guess, I wanted to harvest the forest myself, exploiting the new resource for sentences and paragraphs—acting out as a writer the same paradoxes of use I kept discovering in my reading and interviews.

The research project that has entirely redesigned the hill behind my house, for example, is entitled "Comparisons of Terrestrial Vertebrate Communities and Tree Regeneration Among Three Silvicultural Systems in the East-Central Coast Range, Oregon." Its authors are John Tappeiner, a tall, bearded silviculturalist in his mid-fifties, and Bill McComb, a thirty-something wildlife biologist, deep-voiced and intense. "Examining scales, intensities, distribution and frequencies of disturbances that once occurred in unmanaged forests," they write, "can provide a basis for designing silvicultural prescriptions that produce a landscape that in structure and composition may imitate 'natural' landscapes." Trees fall over in the wind. Trees die from disease. Small fires break out at intervals. All these create "disturbances" or "gaps" in a naturally developing forest, and this may be the key to managing forests for both timber and other "values." Plants and animals apparently survive such small-scale disturbances. A number of small clearcuts can create similar gaps—chunks of timber taken out and used—while still leaving "a matrix of mature forest" to provide suitable habitat for certain kinds of wildlife. Large-scale clear-cutting has the opposite effect, according to Tappeiner and McComb, isolating small islands of trees in a sea of disturbance, replicating rarer catastrophic wildfires. Research suggests that many birds and mammals suffer in these big openings, and there's evidence that new trees won't keep growing back rotation after rotation.

Talking with McComb over coffee one day on campus, the murmur and clatter of the commons all around us, I was struck by how academic our conversation was, how abstract and even literary. The forest is like a poem to him, a complex text whose levels he reads. But his and Tappeiner's interpretive metaphors translate into the falling of real trees, the opening of real gaps. Their "licor measurements of percent sky" have determined the amount of twilight I see as I walk the logging roads and trails, the percentage of sunrise. Their use of a "snag recruitment simulator" has left sawed-off trunks where flickers and woodpeckers are beginning to nest. Tappeiner and McComb write the forest, not just write *about* it; their paragraph indentations are new openings in the trees.

Now there are three kinds of cuts on the hill. At the end of one logging road where I often walk is a substantial clearcut with about a dozen large snags distributed across it. That's the experimental control. To the north and west are "two-story" or "shelterwood" stands, pieces where all but a third of the tall trees have been taken off. And then folded into the rest of the forest are dozens of half-acre to acre "patchcuts," the miniclearcuts that most interest Tappeiner and McComb, replicating blowdown and disease. What are the effects of these different cuttings on mammals and birds? How successful is reforestation in these different openings—what will grow in the partial shade, and how well?

Four-thirty one July morning I went out into the patchcuts with Carol Chambers, one of McComb's graduate students, to help chart the movements of diurnal songbirds. She's a slim, soft-spoken woman from Kentucky, a long blonde braid hanging down the back of her plaid shirt. We would go to a blue- or pink-ribboned stake, sit down in the bedstraw and candy flower, and listen. The forest was alive with bird song, a cacophony of dozens of species calling out, announcing their presence, repeating themselves. Carol would point and say, quietly, "olive-sided flycatcher" or "hermit warbler" or "Swainson's thrush," recording acronyms in a notebook grid and marking approximate distances on a circular target. It was a wonderful morning, the sun coming up orange through the tree trunks, a thousand blended notes in the air. Another day we walked up and down the hills checking live traps for mice and voles. Carol would reach in with a gloved hand, take the animal—usually a deer mouse—by the tail, attach the tail to a pencil-sized scale (the mouse struggling upside down for a minute, its little legs scrambling), clip a toe for marking, and let the specimen go. We talked about nature writing as we went from site to site, sharing our enthusiasm for Annie Dillard and Wendell Berry.

The warm August day I watched part of the logging operation a graduate student from Forest Engineering was doing time and motion studies on the head faller, clocking how long it took him to angle the trees so that they fell clear in the patchcut. Marvin is a precision faller, a friendly and unassuming middle-aged man in suspenders and hard hat. He joked good-naturedly with the research coordinator showing me around, complaining about all the "New Forestry bullshit" and the tricky angles it requires, but you could tell he was proud of his craftsmanship, and he was quick: falling a tree, measuring it, and "bucking" it into two or three pieces in about ten minutes.

Later that day I saw the yarding operation on another site. A high-towered yarder pulled logs to the top of a slope where the swiveling tongs of the "shovel" pinched and whipped them around to a pile. As the "chasers" undid the choker cables and sawed off random limbs, I heard over the sound of their chain saws the foreman shouting about a "serious deflection" problem with the skyline, something about not enough arc between the cable and a ground bulge, a problem sure to be worse on unit ten, next up. My impression was of hugeness and loudness and danger, the swooping and swinging of big logs, though I was told that these 140-year-old trees are nothing compared to the monstrous old-growth the really big machines can yard. The question for the forest engineers on the project is how much this smaller-scale, finesse logging really costs in comparison to conventional clear-cutting (Answer: at least 22 percent more on the first harvest).

Other researchers are involved in the project, too, studying the effects of the cutting on the human community. Becky Johnson, a resource economist, is interested in "the socioeconomic impact of harvesting techniques on residents on the urban fringe." She thinks of the forest, she told me, as a "multiple output commercial asset," and she wants to measure the "noncommodity values" that are a part of that asset, how much people would pay for what they don't currently have to pay for. Questionnaires are going out. One afternoon I participated in an aesthetic-perception survey conducted by Mark Brunson, a graduate student in Forest Resources. He took us to different sites—clearcut, patchcut, old-growth, and so on—and asked us to rate our aesthetic responses on a numerical scale, taking into account sounds, smells, spatial definition, and the possibilities for camping and hiking. We carried the survey form loose as we walked, balancing the sheets awkwardly on our thighs when we wanted to write or using a stump as backing. It was a late autumn day, even the clearcuts soft and full in the fading light.

Not surprisingly, most of us in the study rated the patchcuts as more aesthetically pleasing than clearcuts. From a distance the smaller openings are almost invisible on the more level parts of the hillside. To Mark this suggests that the human habitat can be maintained within the matrix of the mature forest, too, the "viewshed" preserved.

I think of a moment from a tour that summer of the Andrews Research Forest in the Cascades outside Eugene. About thirty of us from OSU's Sustainable Forestry New Paradigm Working Group—foresters and scientists and humanists—were walking single file down a trail through 500-year old-growth, gesturing with our hands, turning around to make points, filling the air with our voices: ". . . reading streams for their level of complexity . . . the role of lichen as nitrogren-fixers in canopy microclimates . . . hierarchical scales . . . complex communities . . . structural diversity."

I think of the plastic ribbons and markers and stakes and trail signs and spray-painted numbers on the trunks of trees everywhere in McDonald Forest, as if the endless studies and reports have begun to show through on the land itself. "Wildlife Tree," one of the latest signs reads, "Please Protect / OSU Research Forests." They're bright red plastic, screwed onto many of the newly created snags, a picture of a pileated woodpecker on the side. And this sentence was tacked to a fir when the harvest began, in the forest behind the house: "The Research Management request that all individuals recreating in the adjacent area please follow these guidelines."

Much of this language infuriates me. Calling a stream "an open-water system," as I heard a hydrologist do the other day, is just silly, and potentially dangerous. The real, concrete particulars get lost in abstractions. If you think of a forest as a multiple-output commercial asset day after day, analyzable only on a spreadsheet, you forget how it smells and what it's like to walk in it in the morning. The important realities are beyond any words. Scientists, of course, don't always use the language of science, don't always view their experience scientifically, but the long-term effects of jargon—like radiation—are sometimes hard to protect against. And sometimes the abstractions are deliberate obscurings. "Treatments" and "prescriptions," after all, are just euphemisms for cutting, for killing. We can't finally trust any of this. Yet at the same time I found myself oddly soothed by the sound of the words I kept hearing, reassured that at least there was some rational method behind what was happening in the forest. Living on the level of abstraction was therapeutic for a while, giving me some perspective and detachment. Deeper than that, hearing the words and models and paradigms over and over again moved me beyond my initial naivete. The language kept showing me that the forest is a complex place, a human place, not just something to look at or find refuge within.

Jeff Garver, the manager of McDonald Forest, is a six-foot-five former track star and Eagle Scout, bearded now and still imposing. I spent two whole days bouncing around in his Chevy Blazer getting the standard tour. He drove left-handed, grabbing for a Big Gulp with his right or talking on one of his two cellular phones, dialing with his thumb. His speech is practiced and unmodulated, its cadences a mixture of law enforcement and public relations, although now and then I'd sense a sharper-edged voice underneath. He kept handing me the keys and making me fumble with the recessed locks on the metal gates blocking the entrances to logging roads.

At the Lewisburg Saddle he looked out at the far view of hills and valley and praised the "gorgeous" plantations we could see from there, their "sharp, tight points" versus the "ratty and beat-up tops" of the narrow strip of old-growth canopy. He noted "the good bird activity" in the clearcut in front of us, and the way the far clearcut, on Forest Peak, blended into the hillside. He designed that clearcut, trying to carve its edges so that they "fit better with the mind's eye." "Straight lines are as ugly as you get. I like the nice rounded contours—like a '51 Ford." Variety, "multiple-use," is what he kept stressing. There were always margins and alternating textures in the scenes he showed me, a quilting of clearcuts and various plantations. In fifty years the same "elements" will be present in these scenes, he said, just in different combinations, as if the forest is a vast temporal playing board, the same squares exchanging position over time.

The patchcuts don't make sense to him, though he's tried to accommodate the scientists. He's doing the best he can in a difficult position, answering to a divided faculty and trying to manage seventy-two separate research projects without the benefit of a coherent long-term management plan. Given public anxiety and involvement, too, he knows he has to practice what another forester I talked with called "sensitive forestry." Still, by the end of our second day, as he seemed to relax more and more, Jeff was calling the New Forestry just "weird science," "deep fungal," the product of computer jockeys and college professors with no real experience working in the woods. He thinks we need to be practical. The brush left by the patchcuts is a fire hazard, there's too much merchantable timber left on the sites, and the trees won't grow back anyway. The goal of the forest should not be research alone but the utilization of the available resources, as in any good commercial operation. Revenue is necessary to sustain the research anyway, and it costs at least a million dollars a year just to keep McDonald-Dunn going.

What Jeff honors is the paradigm of forest management that's dominated the College of Forestry, and all forestry, for two generations. His faith is in the good "site prep" of clearcut ground—poisoning and burning—then the planting of genetically strong seedlings, then judicious thinning over time. "Cut a hundred acres. Plant a hundred acres. Thin a hundred acres. Don't cut more than you grow. You can color that with all sorts of fancy computer models, but it all comes down to this."

But that, of course, is Jeff's model. He, too, reads the forest. He, too, writes it. He sculpts it, changes it over time, putting his ideas into action. His forest is just as complex a place as Tappeiner and McComb's, requiring just as precise a jargon to understand and manipulate (there are "bearing strengths" and "blind leads," "catch points" and "deflection angles," "inslopes" and "tangencies"). What I used to see as simply an expanse of trees, a hillside of fir, is in fact a silvicultural system managed day to day through hard work and insiders' knowledge.

It is a system sustained by a budget, a regular office staff, and up to thirty part-time employees. It is an institution. There are a motor pool and a coffee fund, there are staff meetings and office parties and rows of cabinets preserving old memoranda. Five other foresters—for recreation, reforestation, research, public education, and engineering—branch under Jeff on the organizational chart. There are insiders and outsiders, a pecking order. Each morning people check in at the office, gossip over coffee, return their cups to their pegs, and fan out across the forest to plant seedlings, put up signs, wait for satellites.

The Andrews Forest, coadministered by Oregon State and the Forest Service, is more impressive than McDonald-Dunn. It's only a little bigger than McDonald-Dunn—around 16,000 acres—but it sits in the midst of the Willamette National Forest, miles into the Cascades from Eugene, and it's mostly old-growth, uninterrupted expanses of huge trees set aside for research. It has the old-growth mystique. Walking through its trails you have that sense of being hushed. You're always looking up.

McDonald-Dunn seems small and dull in comparison. It is close to the hubbub of town, an "urban fringe forest" cut up into littler pieces. I've heard it called a "hobby forest," too, significant only because it is convenient. There is very little old-growth except in protected draws where the fires couldn't reach.

And this the single most surprising fact about the forest to me. This is what Bob Zybach taught me that day in June when we startled the napping forester and walked in the ruts of the Applegate Trail: 150 years ago McDonald-Dunn wasn't a forest at all. It was an oak savannah, a prairie extending as far as the eye could see with just a scattering of two or three oak or fir per acre. That's why grass still grows on the forest floor and large stumps are rare. Long branches stick out from the odd big trees in the midst of newer growth, a sign they were once growing in the open, without competition. They're "hooter" trees, savannah oak and fir. Large-scale harvesting wasn't even possible on McDonald-Dunn until the late fifties and early sixties because too few of the trees were big enough. The whole Willamette Valley is the same, the result of "cultural fire"—the seasonal burning practices of the Kalapuya and other Indians.

Early trappers and explorers described the vast expanses of

grass and wildflower, and the smoke that obscured them parts of
the year when the Indians were burning. For instance, we have
these notes from the 1826 journals of David Douglas, a Hudson's Bay botanist who gave his name to the fir:

> 9/27 Country undulating: soil rich, light with beautiful soli-
> tary oaks and pines interspersed through it and must have a
> fine effect, but being burned and not a single blade of grass
> except on the margins of the rivulets to be seen.

> 9/30 (heading south) . . . Most parts of the country burned;
> only on little patches in the valleys and on the flats near the
> low hills that verdure is to be seen.

Other explorers write of choking smoke and the absence of grass
for livestock in the fall, the "grand panorama view of prairie,"
and "the excellent quality of grass abounding" in the spring and
summer. All up and down the valley the Indians would burn,
every season for perhaps thousands of years, altering the landscape to suit the berries and hazel nuts they fed on, harvesting the roasted tar-weed seeds and herding deer into unburned
corners. Even more dramatically than contemporary foresters
and managers, the Kalapuya and other tribes made and remade
the land, "culturing" countless acres of it from Washington to
California.

McDonald-Dunn Forest is the result of "fire suppression" in
the 1840's and 50's when the first settlers arrived, which is to say
that the Kalapuya quickly died out, victims of malaria and smallpox, and that the Douglas fir, after being restrained for so long,
finally claimed the meadows.

Forests are supposed to be old, of course. They're supposed to
be permanent, given. But McDonald Forest is actually quite
young, even ephemeral, when seen from the perspective of biological and geological time, as Zybach explained to me that day
in June, wading ahead of me through the wildflowers. Its Douglas fir are technically "invaders." What I took as solid suddenly
seemed fragile, the trees like feathers, like false-fronts.

"Trees are cheap," Bob says. "They're everywhere." Too many in one spot make him nervous, since he's spent most of his life working beneath them. "Do you know how many people get killed by falling trees every year?" he asks, laughing. What he loves are the sweeping vistas, the tall grasses, the wildflowers remaining from the indigenous prairie. If he had his way he'd clear-cut and burn a big part of the forest, returning it to savannah. That would be aesthetically pleasing to him. That would be restoring the forest to its healthy, "natural" state, fire cleansing the forest of pests and undergrowth while returning nutrients to the soil. There were few significant snags or islands of fir in the forest before the settlers came. The Tappeiner and McComb patchcuts, their "New Forestry" snag distributions, are about as natural as a "garden," just another example of "college sense," the "weird shit" of overgrown "college boys."

Bob Zybach is a small, muscular man in his early forties, sandy hair falling to his shoulders, a former logger and private reforester with a passion for history and a contempt for academics and bureaucrats. He's a "taxpayer" more than a "tuition payer," he told me, even though he's been studying forestry at Oregon State since his business failed in the timber bust of the early 80's. A self-taught historian, lifelong student of Indians and pioneers, he was recently commissioned by the dean to compile a "Cultural Resources Inventory" of McDonald Forest. I asked him to show me some of the sites he'd catalogued. When I picked him up the first time he was wearing a "Save the Rain Forest" T-shirt, the second time a "Desert Storm" trucker's cap. He gets carried away when he talks, acknowledges that he can seem "abrasive." He says he's used to shouting at people over chain saws.

It was great fun bombing around the forest with Bob in the old Buick, debating for hours about what's natural and why that matters, what's really true and what's the product of academic self-interest, the money of the funding agencies. Off Homestead Road we gathered shards of blue crockery from the site of the old Tortora place, first settled in the 1870's. We found the rusted body of a stove reservoir there, too, a remnant from the time of World War I. Two sixty-year-old firs grew from a ten-by-fifteen hole left by the original foundation, the sides rounded now from age, and there were century-old pear and apple trees in the pasture. The rest of the homestead, what used to be oak savannah, is covered now by second-growth fir, a little forest stretching down to the bottom of the draw.

It was there, Bob claimed, in the mid-seventies, just down the hill, that Eric Forsman conducted his first spotted owl experiments as an Oregon State graduate student, coaxing the birds from the trees with mice. Consider that, Bob repeated: catching spotted owls—the symbol of old-growth—near the site of a homestead where old-growth has never been recorded, in a young forest rising from the last of a prairie long ago settled by pioneers.

Later we fought our way through vine maple and alder and dense forest to the site of the old Coote sawmill, dating from the 1930's. We passed a giant cottonwood and a very large maple. Then up ahead, sticking out of the creek like a giant rusted fin, was the perpendicular windshield frame of a Model T. Coming closer we could see the circular hole in the back of the cab where the rear window used to be. Near fallen trunks on the other side were scattered fire bricks from a kiln, the remains of a metal water tank, and indentations in the ground indicating a road and a wide staging area. All around us was deep forest, the wildest we had seen that day. "Some industrial site," Bob said, gesturing.

On the way home, by the side of the highway, we stopped and found several "bearing oaks" or "witness oaks" from the 1850's, their trunks blazed with an axe. Early settlers used these trees as fixed points to find their bearings for the first land surveys. One was inscribed with several small circles arranged in an arc. Not far away, in the same stand, were modern bearing trees, precise survey numbers marked on metal plates.

Because the valley floor was often flooded and marshy, the

hills and ridges of what is now McDonald-Dunn Forest have always been the site of human habitation and culture. That is why the Kalapuya lived here, following the ridge lines to look out at their fires. That is why the California Trail of the 1820's followed the wide bench against the hillsides, taking Hudson's Bay pack trains north and south through the valley—why the Applegate Trail of the 1850's hugged the hills farther down, near what is now Highway 99, taking settlers to their promised farmlands. Jedediah Smith, the famous mountain man, traveled the California Trail, right behind where my house now stands. Peter Burnett, the first governor of California, the first leader of a wagon train on the Oregon Trail, and my wife's great-great-great-grand uncle, passed by Corvallis on his way south to the Gold Rush. My house was built on what was once the Donation Land Claim of a man named Fuller; we live on his upper pasture. Next to us was the claim of Thomas Reed, the first settler in this part of the valley. His house was framed in 1853 by Bushrod Wilson, a well-known local carpenter, and it became a famous wayside for travelers on the Applegate.

For thousands of years McDonald-Dunn Forest has been crisscrossed and carved out and built on, layer after layer of culture sooner or later sifting down to the forest floor. Its cultural value far outweighs the value of its timber in Bob's mind, timber which exists in the first place precisely because of human intervention. For him the forest is not a mosaic but a "time machine," the past lives of its people recorded in vegetation patterns and old orchards merging now into fir, in hidden wells and pieces of tin, in arrowheads and fragments of flint—lives of ordinary people more admirable than the conspicuously consuming yuppies who now live on the forest edge. He imagines the people of the past living in harmony with the land, quiet and slow, wiser than any computer-generated model.

Walking into Bob's rented house near campus you see piles of old journals and documents on the couch and the coffee table and filling up the corners, maps rolled up everywhere or spread

out on the floor and in the kitchen, old history books and tran-
scribed tape recordings spilling out of file cabinets, the blurred
mimeographs of family histories. Here are Alexander McLeod's
Hudson's Bay journals describing the climate and vegetation of
the valley, together with the notes of the Wilkes expedition of
1841. Here are the memoirs of Sarah Cummins:

> Sitting alone and glancing over my past life, long and event-
> ful as it has been, I recall many of its scenes of pioneer adven-
> ture that were marvelous manifestations of the power and
> goodness of God in protecting us in our travels through wild
> regions, inhabited by savages and the haunt of wild beasts.

Here the day book of Lester Hulin:

> to day 5 of us laid in the bushes to watch for indians we heard
> them halloo but they kept at a proper distance we think they
> saw us go in the willows our caravan moved on to a lake,
> then about 3 ms up it and camped distance about 10 ms

> passed around a large swamp filled with ducks geese and
> cranes then passing a good spring we came to a lake watered
> our cattle and passed on over stony roads and at last camped
> without water good grass in sight of another lake distance
> about 14 ms

And Bob is in the process of compiling an oral history of the
Soap Creek Valley, interviewing elderly residents who remember
stories of what the forest was like at the turn of the century, and
before:

> You asked about stories on the trail. I only remember one
> that they told. They come up just looking over a ridge and
> here was a bunch of stuff waving on ahead. They were ready
> to group up, and then discovered it was willows instead of a
> bunch of Indians.

> It snowed and snowed and snowed in 83. Dad used to talk
> about that. He talked about finally it quit snowing and they

decided they wanted to hike over to some friends or family or something for a visit and he said that that old snow was piled up and kind of slippery on the surface. And Dad was walking with a stick with a nail in the end of it. He got down on the side of a hill and he started slipping and sliding down, he said his mother was standing out there shouting "Jim, Jim, Jim, Jim Jim!"

A forest of voices, of stories.

The Thomas Reed house is still standing, near the entrance to Peavy Arboretum. It's a large yellow house now, expanded over the years. A riding mower and a Rototiller were parked in back, near the garage, the day Bob and I walked up the driveway. Through the front window I could see a microwave on refinished kitchen counters. We knocked on the door, but no one was home. Log trucks geared down Highway 99 behind us.

Driving my daughter to her piano lesson last week, to the top of Vineyard Mountain, I counted six minivans, three BMW's, three mountain bikers, and two pairs of white-shorted joggers. Upscale neighborhoods rim the forest, developments with names like McDonald Forest Estates and Timber Hill and Skyline West. Long, split-level houses are built into hillsides, hidden in oak and fir, their cedar decks offering views of the valley and the Cascades. Right now a developer is grubbing out a "real estate cut" at the base of Vineyard Mountain, removing the fir and leaving the madrone in preparation for a forty-eight unit housing development. I can hear the Caterpillars powering from here, over the whine of my computer fan. This morning's paper describes a neighborhood protest at the Benton County Planning Commission last night, over a hundred homeowners expressing their worry that still another development on the mountain will severely deplete the already marginal supply of drinking water.

Over 35,000 "recreation days" are spent in the forest each year, according to a recent Master's thesis. There are mountain bikers and equestrians and hikers on the roads and trails. The Timberhill Harriers run here every weekend. Once we stumbled into a timber carnival over at Cronemiller Lake. High-school timber clubs from across the state were competing in axe throwing and timber cruising and tree climbing. Trucks and campers were parked everywhere. A hotdog stand was set up. Another day we came across a mountain bike competition, the finish line at the lake. Mud-spattered riders came whizzing off the hill, numbers flapping on their jerseys.

Once I was walking with the kids on the 510 Road when I heard jazz guitar seeping through the trees, then lounge music piano. Farther on we saw a wedding reception at Peavy Lodge, tuxedoed young men parking their cars, women hiking up their gowns to climb the steps. We could see a long sheet cake, balloons flying from folding chairs.

It all seems natural to me now. There has always been a "human/forest interface" here, beginning at least as early as the first makeshift shelters of the Kalapuya.

Perhaps the central effect of my studying the forest this last year has been to complicate my understanding of the "natural." On the one hand some of the rhetoric of environmentalism seems naive and unconsidered to me now, even foolishly arrogant. We can't ground our arguments for what is right on some sentimental longing for the unspoiled, as if only what is nonhuman is good. That's to be ignorant of history, as well as to misunderstand our own responsibilities. Wendell Berry argues that just as culture depends on wilderness—just as we need to preserve wilderness to survive as a culture, spiritually and physically— wilderness now depends on culture. Setting aside the Andrews Forest to remain as old-growth is a cultural act. It is the drawing of a line, the creating of value. Preserving wilderness means erecting fences, fumbling with locks in metal gates.

But that argument also works another way. Often in my conversations with foresters and scientists a policy or practice would be explained to me as if it were inevitable, as if things had to be done that way, inexorably. Sometimes there was an arrogant privileging of expertise, an invoking of the tropes of objectivity and practicality, as if my concerns with aesthetics and spiritual values were merely subjective, merely personal. But the history of the forest argues something far different. It argues for change, for patterns too shifting and evolving to justify any single practice or claim of ownership. Clear-cutting isn't an ancient, inevitable method: the first settlers didn't have enough trees to cut; selective logging was practiced between the wars; Mary McDonald started giving money to the School of Forestry for the expressed purpose of encouraging reforestation, not harvesting.

We can do anything we want. As a nation we can choose, for example, to pay more for timber. Who's to say that the current price is inevitable, objectively right, that there aren't other values we might pay for other ways? On McDonald-Dunn Forest we can choose to pay a recreational use fee, as some have suggested, to take the financial pressures off the forest. We can decide, as a community, that the first goal of the forest should be education and research, not the generation of revenue—or we can choose to clear-cut and burn all 12,000 acres. Science imposes limits of fact. Trees grow at certain rates in certain soils in certain climates. Ecosystems function according to complex interchanges of energy. But even then these are facts to be interpreted, the basis of policies we need to construct. The history of the forest shows that it has always been cultured, shaped. It has been made. No policy can be justified on the grounds that it is pure.

One evening at the end of the summer I walked to my first meeting of the McDonald Forest Trails Committee, over in the Forestry Club Cabin, a mile or so from the house. It was odd to be walking through the forest to a committee meeting instead of driving to town for one. It almost seemed as if time had fallen away and I was hiking to a gathering at the Reed place for the evening, to catch up on the news and tell stories with my neighbors.

The meeting was a potluck, pasta salad shining under the cabin's new fluorescent lights. The room smelled like a school cafeteria. The agenda was up on a marker board in the front, items listed in ten-minute increments, and by the time I got there discussion of the "Multiple Use Trails Map and Guide" was underway. Representatives of the equestrians were politely complaining to representatives of the mountain bikers while Mary Rellergert, the recreation forester, practiced her conflict-resolution skills. Everyone kept using the term "user group": how do we accommodate the overlapping concerns of these different "user groups"? how can we get the full participation of this or that "user group"?

Two hours later I escaped back into the dark, familiar forest, to the sound of crickets and the smell of smoke from the field burning in the valley. It was early September, still warm, exactly a year since the loggers began their work. The last of the sun glowed over the edge of the shelterwood.

I could have named the different trees I brushed past, explained the theory behind the shelterwood, pointed out where the California Trail came in from the south, following the 510 Road. But I was glad to be walking in the dark, with just the sound of the trees around me. I was glad that the loggers were gone, the forest returned to its own rhythms and silences. After a year of studying and learning and interviewing experts, it's still the surfaces I value, still the feel of things, the smell of blackberry and needle duff, the play of shadow. McDonald Forest may not be as spectacular as the Andrews, but living on and near it through the seasons I have come to feel for it a special affection.

And this is what I want to argue for: for local knowledge, for personal knowledge. After all the terms and ideas and paradigms, what I value most is the sense of familiar ground, of a place I know well enough to find my way home in the dark, and I want to argue for that, just for the feeling of being here, fully, with the heart and the senses. I want to argue for the mind at rest.

The difference is that I know now I will have to *argue* for these things. I will have to fight for them and represent them publicly, and to do that I will have to know something about current silvicultural practices and the ecology of the forest. I will have to attend more potluck meetings at the forestry cabin. To make possible for others the knowing of a forest by heart, I will have to learn to speak the language of "user groups" and "recreationists" and even "multiple-use managers." I will have to make my own voice heard in the forest of voices.

Life on the Edge

All ethics rest upon a single

premise: that the individual is a

member of a community of

interdependent parts.

—Aldo Leopold

"Cultivate simplicity, Coleridge," Charles Lamb wrote his romantic friend in 1794. "Or rather, banish elaborateness; for simplicity springs spontaneous from the heart, and carries into daylight its own modest buds and genuine, sweet, and clear flowers of expression."

I spent two happy weeks spading my new garden by hand, chopping up weeds and hard clay, pulling out roots. From the angle of the garden plot I had a good view of clouds rolling over the top of the house (this is looking south), the weather patterns changing. It's a quiet, fragrant spot, half ringed by trees.

But then I sat down to read about gardening, in preparation for planting. In the first chapter of Steve Solomen's *Growing Organic Vegetables West of the Cascades*, for example, there are seven charts and two tables with titles like this: "How Field Conditions Affect Germination Percentage" or "Effects of Soil pH on the Availability of Plant Nutrients." Arrows and x/y axes and parabolic curves go every which way, and underneath are lengthy discussions of cellulose and lignins, soil fungi, the structure and composition of earth.

Even if you decide to ignore the charts as unnecessarily scientific, you still need a corresponding practical knowledge every bit as involved and precise. The practical science of peas, for example. My "Territorial Seed Catalogue" instructs me to sow seed 1 inch apart, 1 inch deep, in rows at least 12 inches apart, or broadcast seed on raised beds at 1 lb. per 100 square feet; 5–10 pounds of bonemeal per 100 feet of furrow or 100 square feet of raised bed may greatly improve growth. I'm not to plant non-enation-resistant varieties before mid-March (enation is a virus wilt transmitted by aphids), although the new genetically engineered enation-resistant varieties can be sown throughout the summer.

That all this matters I learned for myself when I planted my Oregon Sugar Pod II *two* inches deep, having misjudged the distance between the middle joint on my index finger and the end of my finger. I waited and waited and then consulted the catalog, reviewed the steps, discovered the misjudged joint, raked off a layer of dirt, waited again, replanted part of the bed—and finally the peas emerged in spots and patches, about half the expected crop, winding and delicate, the small broad leaves folded together like little mittened hands.

It's necessary knowledge I'm talking about—and the disparity between my first green images of gardening and the complexity of actually doing it. You need to learn terms and techniques, practical if not theoretical. You need teachers, either books or people. There's trial and error, a requisite passage of time. There's always a discipline.

In the interest of living a simpler, natural life, I decided to buy a bike and commute to school. I wanted the wind in my face and the sleek thighs of exercise. A trip to the library and a tour of the bike shops gave me my first lesson: a 42-tooth small chain ring, a 52T large ring, and a 13−23 or 13−24T freewheel cluster provide suitable gear selection for hilly terrain; to get the right frame fit, according to *Bicycle Magazine's New Bike Owner's Guide*, I need to factor in inseam length, recommended frame size, recommended bottom bracket height, recommended crank length, torso plus arm length, recommended top tube length, and recommended stem extension. All kinds of brochures displayed all kinds of equipment, glossy page after page, brilliant with hardware.

And there's more than equipment involved. There's the science of pedaling. You don't just hop on the thing and ride away. "Resist the temptation to stomp on the pedals," the *New Bike Owner's Guide* tells us. "Instead, feel and visualize your feet spinning in smooth circles. Position the balls of your feet directly over the pedal axles and choose gears that let you maintain a cadence of 90 revolutions per minute (rpm). You can calculate this by counting the number of times your right leg reaches the top of the pedal stroke in 30 seconds, then multiply by two."

Even subtracting for the silliness of this—the American tendency to make a science (and a business) out of even the simplest activities—there's important information here. The numbers and terms describe a physics you really feel when you're cycling home up a long country highway, against the wind. I paid great attention to the geometry of seats, for example, after waking up unable to move the second week of commuting. I'd carry a wrench in my panniers and stop every mile or so to adjust the tilt and height of the thing—went through three different saddles, increasingly more expensive, before I discovered the joy of the Avocet Gelcap Women's Mountain Bike Saddle A-30, worth three times the thirty dollars I paid for it.

It's not that there hasn't been spontaneity in the garden or on the bike, moments of greenness and gliding and thoughtless movement like I first hoped for. There's a spontaneous delight in technique and terminology, for that matter, once you get past the initial stage of resistance and confusion. Soon it becomes second nature, internalized, and you're proud of knowing what you know.

The point is that naturalness isn't natural. It has to be worked for, prepared for. It comes only in the context of discipline and structure over time. Looking up from the peas or racing down the mountain roads, we experience a freedom only possible because of prior commitments—a momentary freedom, too, a fleeting sense of satisfaction that comes only now and then in the course of the workaday routine.

It's a lesson I always resent, that I keep having to learn, and each time a part of me remains nostalgic for the greenworld I have to leave aside (again). It's the lesson of the clear-cutting, the master story I see repeated now in every part of my life.

I moved the family to the edge of the forest because I wanted us all to live more deliberately. I wanted us to live where we could look at the trees. But then the logging began and the trees started to fall and before long I found myself immersed in the rhetoric of forestry, lost in the jargon of still another discipline, and suddenly the forest was not a forest anymore. It was a silvicultural system. It was an ecosystem. It was, in the working definition of the Society of American Foresters, "a plant association predominantly of trees or other woody vegetation occupying an extensive area of land." In David Smith's *The Practice of Silviculture*, a standard introductory textbook in forestry classes at my university, there are more figures and charts and graphs: "Figure 1-2: the relationship between the period of regeneration and the period of intermediate cuttings for a sequence of even-aged stands managed on a 60 year rotation according to the shelterwood system"; or "Figure 1-3: typical examples of four different kinds of stand structure, showing appearance of stands in vertical cross section and corresponding graphs of diameter distribution in terms of numbers of trees per unit of area."

One Saturday I attended a workshop on "New Perspectives in Forestry"—three hours of sitting on hard chairs and watching slides and following lighted pointers, the air full of important terms: "site productivity," "stand configurations," "the viability of streams as a system."

The forest is not my garden. I can only harvest it aesthetically. The terminology is a necessary tool for those who do manage it, those who struggle to balance the competing needs of timber production and "noncontingent values." I struggle to learn the jargon *because* I don't have control. I struggle to learn because I'm afraid that the managers of the forest see it only as a system, only as an economic unit, because I know that I'll need to argue for the value of what I experience actually living here, on the edge of the trees, on the urban fringe—what I see and smell and hear every day. I learn the new difficult discipline under the beautiful surface because I need to become an advocate for the surface. Simplicity is a responsibility.

There's a willow tree in our front yard, graceful and delicate as willows are, and in the spring we started noticing the browning leaves on the lower branches, a blight creeping toward the top. After a few phone calls and a little research, I climbed into the center fork with pruners and a saw and spent an afternoon clearing out the middle of overlapping branches, trying to open up space so the living branches could breathe. Viewed from above the tangling looked a little like the network of muscles and tendons in the encyclopedia plates of the body, though more gnarled and random.

The kids played at the bottom, weaving the willow wands into helmets and garters, chasing each other with whips, finally dragging the branches to the top of the hill and making a little lean-to against the side of a fir. From the middle of the willow, wedged in and trying to hold the saw, I could only hear them, their laughter and shouting. I was too busy banishing the elaborateness of the wilting tree, trying to cultivate its green simplicity.

(A talk given to an introductory forestry class one October morning at the foot of the clearcut)

I appreciate Mark including me in the class today, asking me to come and say a few things on behalf of the homeowners who live on the edge of the forest—that species affected by the changes in the habitat. I'm glad to talk about aesthetics and other intangible values implicit in the "New Forestry," since it was aesthetics that brought me to the woods.

I live not far from here, about fifteen minutes by foot from that direction [pointing north]. Max and I walk the 510 Road two or three times a week, in fact [he's whining, straining at the leash to sniff the students circled around us]. There's a cutoff trail just about a hundred yards above the house, and we often walk up the 510 Road to this clearcut and back. It's a good two miles, and flat.

Walking here today, thinking about what I wanted to say to you, I realized again how little the harvest bothers me anymore. It was over a year ago now the loggers started in, and I grieved and panicked and wrote letters and attended meetings, and I still wish, very much, that the forest was the way it was before the harvest began. But I have to admit—and this is a kind of confession, I guess, given some of the things I've written and said—I have to admit that this morning, as Max and I walked our familiar route, and the clouds kept moving through the trees and changing shape, and the forest kept opening up before us, I was feeling much of my usual solace. I'm a little embarrassed to say this, but the woods are still peaceful and beautiful to me, they still satisfy me deeply, and I've been trying to figure out why. Why hasn't the cutting entirely ruined the forest for me? What makes a landscape aesthetically pleasing?

Right now I can think of ten factors.

(1)   Familiarity and Routine

I've walked this road so often I've become comfortable with it, at home, despite the new cutting. It's not just that I'm accustomed to the gaps. It's that walking in the forest through so many moods and seasons I've developed an intimacy with the landscape, both gaps and trees. I know it the way I know a page I've pored over, word for word. Layers keep revealing themselves: a flicker's nest, a deer trail, the broad sweep of the hill.

Other times I'm walking here and I know the forest well enough for it to be neutral, no impediment to my thoughts. I don't have to find my way or see things for the first time and so am free to let my thoughts go, to let a rhythm of thinking and remembering establish itself, and the landscape fades, then, can be taken for granted (though all the while there's this sense of it, still, of its hardness and presence, of the automatic earth my feet keep stepping on, the background of trees: both awareness and forgetfulness, in a balance).

(2)   The Absence of People

Sometimes I think my main requirement in a landscape is the absence of people and chattering and Pepsi cans and mini-vans. Until today Max and I have walked here for months without seeing more than two or three other walkers. Evidence of the human is everywhere, I guess, in the stumps and the flags and the spray-painted numbers, but even so the harvested, managed forest is on the other end of the scale from the shopping mall built up the street from our old house in town (over everybody's protest)—far superior to the new housing development being carved out farther down the mountain. I much prefer stumps and new seedlings and the bare hills to asphalt and foundations and heavy equipment. I'll take any clearcut over any McDonald's. What the spirit craves, I think, is its own absence. It longs for the not-human, for the plain fact of things, for the slow growth of leaves, the obvious trunks and branches, the flashing of birds.

I grow so tired of my own voice, so often distort or contaminate what I make or propose, that what I most need is a landscape I didn't invent, some otherness I can't be blamed for and that won't immediately yield to me. We crave forgetfulness.

(3)   Spatial Definition
(Or the Aesthetic Success of the Patchcutting)

This strategy of checkerboarding little one-acre clearcuts is meant in part to be an aesthetic compromise, a way of extracting some trees without devastating the forest as a whole, and I think it partly works. There are holes in the forest now, small vistas opening up within the larger structure of the trees, but the larger structure of trees still exists, the patterning of trunks and branches extending off into the distance. It's still a forest. Walking through it you still generally think: trees.

Psychologists talk about "spatial definition" as a key element in the appreciation of a landscape. We like what we can distinguish from other things, see in relation—foregrounds and backgrounds, perspectives. Photographs of open plains or dense forests always score lower in the tests of random participants than scenes of mixed elements, more open forests, fields with houses and barns or small hills. Apparently we have some ancestral memory of getting lost and so an instinctive need for landmarks.

(4)   The Solace of Open Spaces

In fact, I kind of like some of the openings the patchcuts create. They let in more sun. There's more sky, more horizon: stars at night, clouds and distant mountains by day. Many of the patchcuts look ragged and junky glimpsed in a row, but sometimes, coming around a corner on this road, you have a view of several interrelated patchcuts that seems like an opening up, an unfolding. Sometimes you get the sense of perspective, interlocking rooms leading your eye off into the distance. There's a pleasing tension between openness and fullness, a suggestion of depth.

Frankly, I don't even mind this clearcut itself [gesturing toward the bare hill behind the class, forty acres or so—the experimental control for the patchcutting]. I wouldn't want the whole forest clear-cut, but I have to admit that I find something actually pleasing about passing through trees into a broad clearing. It's the visual equivalent of breaking through. There's a cleanness and spareness about the view, a quality of sweeping up, of expansion, not to mention the panorama of the Cascades just visible over the tops of the lower fir. Clouds pass overhead, varieties of weather and light. You can see the rest of the forest, have a perspective on that fullness not possible from inside it.

I grew up in Eastern Washington, desert and wheat fields and the Palouse all around me, and I often long for that landscape. In the winter, especially, the valley can make me feel claustrophobic, and I want nothing more than to get in the car and drive up the gorge until I come to the wide open country again.

Who's to say that a plain is less beautiful than a forest?

(5)   Acquaintance with the People

I've spent some time now with the foresters who manage this place, shaken their hands, had coffee and eaten lunch with them, walked with them in the woods, and seeing their faces, hearing their voices, I've been sure of their commitment to the forest, their care and concern. They seem like good people to me, with ideas and paradigms and training and careers of hard, careful work, with expertises that make sense. I've glimpsed a young wildlife biologist jogging on the Powder House Trail, seen a forest ecologist at the store, picking out bananas. The cutting has been humanized. I can't walk up here or look at the gaps in the forest without thinking of the sincerity and craft of the people behind the cutting.

(6)   A Sense of How the Forest Is Managed

I know now from having heard the talks and looked at the maps that the foresters who manage this place think of it, minimally, as a mosaic. When I was shown around I was always shown views that contained a number of different elements and ages of trees, a clearcut here, then a fifteen-year-old stand, then a thirty, a fifty, some hundred-year-old, and even some old-growth. What I've been told, what makes sense to me, is that for economic reasons alone the managers can't afford to clear-cut anymore than they have. There's not enough land to sustain that level of harvest.

Others in the College of Forestry see this mosaic management approach as too sloppy and uncoordinated, arguing for a more coherent plan. That seems right to me, too, just as an outsider.

But I'm reassured that even minimally the forest has to be kept varied and in pieces, that in any single far view there will always be a large number of trees. I keep thinking of something Jeff Garver, the forest manager, told me on the other side of the mountain here, as we were looking out over Lewisburg Saddle into Soap Creek Valley, the checkerboard of cuts and trees spreading out as far as we could see. In a hundred years, he said, all these same elements will be here, in this scene, just in a different combination, a different arrangement—the clearcut a hundred-year-old stand, the hundred-year-old stand a clearcut.

(7)   A Layperson's Understanding of Theory

I've heard the jargon and seen the graphs and understood the basic reasons for the research they're doing here, what they hope to study—the costs of smaller-scale harvesting, the effects on birds and wildlife, the success of reforestation—and the conceptual coherence of all that has reassured me. In part I've let myself be soothed by euphemistic jargon. Calling a cutting a "harvest" or a "treatment" softens the violence a little, abstracting from the harsh realities, and there's a way to let that softening work on you without being deceived by it. But part of the relief, too, is knowing that the cutting isn't irrational or impulsive or motivated by profits only. Behind the new gaps in the forest are clear, coherent, well-reasoned paragraphs in scientific proposals. Behind them are untested hypotheses and important research goals, goals I completely endorse. After all, the cutting done here has been done in the interest of exploring alternatives to clear-cutting, exploring ways of harvesting that also take plants and animals and humans into account. And a community of scientists is thinking about all this, and there's a tradition of theory and publication to support their discussions, institutions and grants and professional organizations to certify any actual work that might get done, and the process is slow, considered, and recursive. On several levels of abstraction, in other words, the cuttings make sense, and that makes them aesthetically more acceptable. Theory orders them, and order is pleasing. There's a purpose to the cutting, and purpose is pleasing.

Knowledge of the theory informs the eye, helping me see things I wouldn't have seen before: the deliberately engineered snags or "wildlife trees," for example, or the seed traps, framed screens left on stumps to catch the naturally distributed fir seed. Theory opens up the landscape, reveals new details to take pleasure in.

(8)   A Greater Awareness of the Complexities and Contradictions—a Greater Sense of Confusion

The cutting in this forest has led me to think more about the timber crisis in general. I take all the articles in the paper personally. Values compete—the aesthetic and the economic, the preservation of wilderness and the necessary production of timber—and now, after the cutting here, I'm just more aware of all the dilemmas, feel them more immediately, and that changes how I view the forest. I'm not able to see the trees innocently or naively anymore.

I know that a timber worker might think of the clearcut as beautiful, representing work and family. I know that a reforester might see the uniform canopy of a plantation forest as beautiful and an ecologist regard the same canopy as a sign of ecological barrenness, something a blue jay would have to pack a lunch to fly over. Two people, depending on their jobs and their education, might experience a stand of old-growth as either awe-inspiring or an angering waste of wood fibre.

Today the paper reports that the Diamond B Lumber Company in Philomath is closing its mill, laying off 148 workers. The owners blame environmentalists for the absence of timber. "We're just guys trying to make a living," one of the laid-off workers is quoted as saying. "It's the environmentalists' right to believe the whole world should be a park, but I don't have to believe that." In their oversimplifying and stereotyping and superficial thinking, statements like this always make me angry. But they also make me stop and think. They express legitimate struggle and paradox, too. I keep thinking of them as I walk through these woods, on this road, watching the clouds move up the hill, and that complicates my aesthetic judgment, suspends it.

A colleague in Political Science has done a survey of Oregon attitudes about the environment, comparing them to attitudes nationally. Oregonians are less sure of things. In response to the question "Should clear-cutting be banned?" 44 percent of participants nationally "strongly agreed" but only 35 percent in Oregon. Should more wilderness areas be established? Forty-eight-point-three percent of the people in the nation strongly agree, 25 percent in Oregon. People who live close to trees, are dependent upon them in whatever way, are less convinced about any generic policy or blanket statements.

(9)   Bird Song

One result of the reading and research I've done since the cutting started is an awareness of all the birds in the forest—juncos and chickadees and warblers and thrushes and hawks and owls. A pair of pileated woodpeckers, old-growth indicators, are regular visitors in my backyard. I'm starting to be able to identify bird songs. I keep track of what species come to the feeder on the deck. Walking this road now I have a new source of pleasure, partly compensating for the missing trees, or at least I'm noticing now dimensions of the forest that were always there, available to imagination and the senses, unrecognized. The forest exists in its fullness and depth at more than one level at once, in more than one dimension.

(10)   Temporal Seeing

Foresters are always thinking ahead, looking at any present scene partly with the eye of the future. When a forester looks at a clearcut he sees simultaneously a ten-year-old stand blocking the view of the valley. When he looks at a full stretch of trees he imagines the thinning and then the falling twenty and eighty years hence. The imagined future is a filter blurring the aesthetics of the present.

Even in my short time I've seen the prospects change with the seasons and the light. I've seen the clearcut raw and scarred right after the loggers left, in February, no green leaves to hide the stumps and tire tracks. I look at it now in the fall, less than a year later, and the vine maple are turning red and the sun is just now coming out and the fullness of shadow is softening the hill (the sun flooding the hillside suddenly, casting low shadows, and the movement of clouds above that, and blue showing). I know that in the spring new trees will be planted, and that in ten years those trees will be ten feet tall, and in twenty years taller still. Sometimes for a moment I can imagine the ground as a forest, and that image changes my perception of the present barrenness, suspends and complicates it.

To see the forest temporally is to see it with patience. To see the forest temporally is to see it with some measure of trust and acceptance. It's to make a commitment to seeing it again and again, a commitment to coming back day after day and studying it, reading it inch by inch, with discipline and respect, over years. To see the forest temporally is to see change and growth and the cycle of things as beautiful, not just the single, static scene.

I don't mean that I've attained that kind of patience or acceptance. I remain resistant to change and suspicious of the powers that be, though more complexly so.

But I have glimpsed the temporal, at least, know it as that crucial missing layer in my seeing, and my life, a new unimagined dimension, like bird song, only deeper, reaching farther. I know that there is a cycle and there is change and nature is never still.

We shouldn't be chauvinistic about the present, whatever its urgencies and attractions.

Wordsworth said we "half perceive and half create" the beauty we find in nature, and I guess that's what I'm saying, too. On the one hand I see the forest differently now because of the ideas and knowledge I bring to it. I accept it because I know more and am more confused about notions I used to take for granted. On the other hand my sense of this landscape, my pleasure in it, seems to come from the outside, too, unbidden and uncontrolled, surprising. I didn't expect to feel the way I do and even resisted it. Satisfaction is the feeling actually produced in me when I walk here over time. Enjoyment is what I experience in the presence of these trees and these openings, just empirically, prior to thoughts and theories.

Don't misunderstand me. When I walk up in the old-growth on the other side of this ridge [pointing southeast now] my visceral response, unbidden and uncontrolled, is a feeling of deeper respect and reverence, deeper quiet, than any satisfaction I feel on the 510 Road. The depths are greater, the shadows fuller, both the sweeping up and the sense of enclosure more profound. The emotional and sensory reactions the old-growth produces in me are stronger, in short, than the reactions produced on the 510 Road, and they seem to have less to do with my own imagination—seem the product of the big trees themselves, seem in fact a silencing of thought.

I don't think I've been taken in. Temporal seeing also means constantly being on the lookout for compromises and backsliding and discrepancies between what's said and what's done. The forest still lacks a long-range management plan, and it's not at all clear how committed the management really is to citizen involvement and the "sensitive forestry" showcased here. The current plan for this stretch of ground is to come back in ten and then twenty years to remove the remaining two-thirds of the trees, and I'm naturally concerned about that, too, both the aesthetic and the ecological effects. Will the replanted trees grow and what will they look like? What will the rest of the forest look like, the rest of the 12,000 acres, beyond this particular hillside? What will the financial and political pressures be in the future? Are we being told the truth, or all that we need to know? If anything I'd like to see myself as one of those pesky, persistent, letter-writing and meeting-attending laymen every institution needs to prod and check it. The price of landscape is eternal vigilance.

All I'm saying is that these things are complicated. All I'm saying is that this landscape, too, this patchcut, worked-over, altered forest, has its own pleasures and compensations. Landscape aestheticians and psychologists talk about "affordance." We deem something beautiful because of some ancestral memory of what the landscape "affords" us, what it enables us to do, which is why forested scenes are always the most highly rated in the surveys. A forest (our instinct tells us) affords shelter and material and food and landmarks and so we are attracted to its depths and vistas. We say it's beautiful. What this particular stretch of forest affords me is an opportunity to walk with my dog and smell the wet ground and look off into distances, and for that I'm grateful. It affords me a chance for intimacy with trees and rocks and natural things, and in that sense, with the voles and the deer mice, apparently, the chickadees and the flickers, I can attest that the patchcutting has not completely destroyed the habitat I need. So far at least, there are still enough trees.

Professor Jim Boyle
College of Forestry

Dear Jim:

It's interesting how people are reacting to some of the things I've been writing and saying—how they often don't actually read what I've written, see the actual words, but immediately make assumptions, start stereotyping me. Buttons seem easy to push lately. When I suggest, for example, that it's time for people like me to start taking more responsibility for where and how they live, to stop trying to escape, I don't mean that I'm therefore going to chain myself to trees, that I'm opposed to any cutting of trees or even opposed to the cutting of the trees behind my house. I'm only saying that I can't ignore the complexities of the issues. I'm saying I want to learn—learn, among other things, about the economies of this situation, all those intricacies.

I'm simply saying that the aesthetic and the personal matter—I'm trying to describe that dimension of things. It doesn't follow that I'm therefore against harvesting of any sort. In fact, my point is usually not to make an argument at all but to share an experience.

I wonder if there's something in the act of sharing experience, in the form of story, at odds with cut-and-run commercial forestry and its various slogans, fact sheets, and apologies.

Yesterday you got me to thinking in particular about the old argument that "people who live in wooden houses shouldn't protest clear-cutting" (apparently the gist of the graffiti about me in the men's room at the College of Forestry). You got me mad, I guess, or the graffiti you told me about got me mad, and that led me to thinking in terms of arguments myself, sorting through the premises and the logic.

—Saying that wooden-house owners shouldn't worry about cutting is like saying people who eat shouldn't worry about overeating, shouldn't try to have a healthy diet—or that people who drink a glass of wine with dinner shouldn't be opposed to drinking and driving. It's to equate all use with wasteful, excessive use—with unlimited use.

—It's to make an either/or argument, to polarize the issue, to make it all or nothing: either massive cutting and exploitation, or no houses. Surely there's an in-between.

—It's to assume that people like me think *no* cutting should happen (when that's not our position).

—It's to assume that efforts at conservation or preservation are solely or largely responsible for the decreasing timber supply in the first place, when the factors are far more complex (as you know): waste, for example, or automation in the mills, or the failure of timber companies to retool for second-growth timber. I heard on the local news last night that 17 percent of all timber harvested in this country goes into making wooden pallettes, pallettes which are used only once and then thrown away. An article in the *New Yorker* last year made another claim that's stuck in my mind: timber jobs declined by 15 percent in Oregon the last decade, at the same time that the amount of wood removed from national forests increased (under the policies of the Reagan administration) by 16 percent, a discrepancy that analysts attribute in part to automation and in part to the export of raw logs to Japan.

—In fact, I'd argue that the main reason for the drop in the timber supply and so the loss of jobs is that we've harvested all the trees. It's not that we've locked up a huge reserve but that we've used it up. There's only 5 percent of the original old-growth left, apparently. Five percent. Cutting it now would only delay the problem a few years.

—The usual argument also assumes that there's only one kind of harvesting, only one way of doing business, when in fact there are several ways of harvesting—patchcutting, for example.

—It assumes that reforestation is successful, trees renewable generation after generation, when that's not at all clear. Many experts worry that plantation forestry can't be sustained for more than a few rotations. (Again: I know you know this.) A recent congressional report, for example, claims that the government has deliberately overestimated reforestation and growth rates on national forests in the Northwest in order to justify higher logging quotas that in fact cannot be sustained. Many of the replantings simply are not growing back, it turns out, the new trees dying because of poor soil, perhaps, or poor genetic stock to begin with or poor management practices once the stand is underway. Only 64 percent of the volume cut in our forests is actually being replaced through reforestation. That's the estimate. Many people are starting to feel that the only way to keep up a supply of timber is to manage forests differently, cutting less and selectively, replicating the natural cycle of disease and fire, letting windfall rot and return nutrients, letting trees grow longer and with less supervision—managing forests so that they look as messy and as complex as old-growth forests, in short. Rethinking clear-cutting may be the only practical thing to do. It may be the only way to maintain production of wood.

—The wooden-house argument assumes that people like me aren't willing to pay for their choices, aren't willing to face the cost of genuinely maintaining multiple values. I'd be willing to pay more for lumber and other wood products (why is the assumption always that the only good material in an economy is *cheap* material, the cheapest possible?). I'd be willing to live in a smaller house. I'd be willing to pay a user-fee for living near and recreating in McDonald Forest.

—I'd even be willing to re-evaluate the way I live in more fundamental ways. I'm not sure, for example, that we can continue living in large, single-family houses or that we can continue living scattered across large areas of land or that we can continue our dependence on cars. Maybe it's time to rethink how our communities are built: multi-family housing, mass transit, etc. In other words, the argument that it's either continued exploitation of resources or the loss of the American way of life doesn't take into account the possibility that "the American way of life" isn't immutable, a given, an undebatable good.

—The wooden-house argument, to put this another way, assumes that there's only one kind of value—"economic" value— and that economic value can only be measured in crude and obvious ways.

—It also seems to assume that all forests and tracts of land should be treated equally, and in particular that all forests should be managed as commercial forests. That just doesn't make sense. Each particular forest or stretch of land has its own character, biologically and in other ways. I'm not arguing that there should be no commercial forests, only that it doesn't make sense for McDonald Forest to be one:

—What's unique about McDonald Forest is its proximity to OSU and Corvallis, and it should be managed to exploit that uniqueness.

—What McDonald Forest can do better than almost any other forest is serve as a convenient teaching and demonstration lab for the College of Forestry. That's the way to get the maximum use out of those 12,000 acres: the educational value far outweighs whatever a commercial operation might yield on that small amount of land; what people might learn is of far more potential value than the small amount of wood fiber more extensive harvest would add to the world's supply.

—That would involve harvesting of trees, of course, but in different ways and for different reasons.

Why treat an apple like an orange? Or a classroom like an assembly line?

I realize that the College of Forestry needs the revenues from harvesting to support its teaching and research programs. The forest funds the College: that's the argument we all keep hearing, over and over again. But there's a circularity to this repeated claim. The assumption seems to be that current teaching and research programs need to be funded at just the same rate forever—that all the research and teaching should be funded— that the purpose and value of these programs, as well as their quantity and cost, are simply given, unavailable for argument and review. As far as I'm concerned, everything is up for grabs. If we decide to fund fewer research projects, we can fund fewer research projects. Research shouldn't jeopardize the integrity of this forest. The revenue needed for programs shouldn't lead to harvesting that would undermine or compromise just those ecological values we ought to be teaching and researching.

Or, to ask the basic question, what ought we to be teaching and researching? Not all research is equal. Let's establish priorities and values for this forest and then relieve pressure on it by excluding all research and harvesting not in keeping with those goals.

Besides, I get the impression sometimes that the teaching-and-research-funding argument is just an excuse for harvesting anyway, or at least that money-making often becomes the overriding motive itself, perhaps even inadvertently. Let's make sure that isn't true. Let's make sure that the needs of teaching and research really do drive the harvesting and not the other way around.

As for the argument that the original Mary McDonald bequest requires harvest of timber for the benefit of the College, I'd say (1) that's subject to interpretation. As I read the language of these letters and other documents, the emphasis is on "research" in some broad and vague way and if anything on "reforestation," not harvest. McDonald's husband was a timber baron, and it's clear from her letters that her desire was to give something back to the state he'd gotten so much from. (2) The models and understandings of the 1920's shouldn't govern policy in the 90's anyway. Things change. Balances need to be righted. New demands and challenges arise.

As for the argument that McDonald Forest "belongs" to the College of Forestry and so is none of my business—that the College can do anything it wants with the forest—I'd say (1) the forest "belongs" to a unit within a state university designed to serve the needs of every citizen in the state. It's part of a public institution, not a private enterprise. (2) One of the clearest insights from the new ecology is that ecological systems transcend property boundaries. We're all downstream. What happens on one tract of land affects all the land in the ecosystem—perhaps land hundreds, even thousands of miles away. The relationships are intricate and far-reaching. A nuclear power plant can't do whatever it deems necessary, regardless of its impact on the land beyond the chain link gates. In forestry there are similar kinds of fallout, just as pervasive, not simply visual and aesthetic but concrete and practical, having to do with erosion, water levels, the integrity of the land.

What really makes me mad, Jim, is the silly, knee-jerk, narrow-minded, unexamined thinking the graffiti reflects. What makes me mad is that the graffiti represents thinking out there everywhere. What makes me mad is that so much debate about natural resources hasn't gotten beyond the level of graffiti.

It seems to me that we won't be able to break through the current gridlock in the management of our forests until our arguments become real arguments, not just slogans and one-liners but careful, reasoned, sincere inferences from fact—arguments based on what we all most value, on what binds us all together. There'll be disagreement in that kind of debate, of course, but that disagreement will be genuine, and it will concern what really matters.

Sincerely,
Chris Anderson

IV

Mowing the lawn has always given me a private pleasure. There is the smell of gasoline and the roar of the mower, yet there is also the smell of the cut grass, and the sun beating down. The clean, even rows give the illusion of order and accomplishment. There is pattern and regularity, and the feel of the earth beneath my feet—the sense of covering ground, my own ground. Or maybe it's harvest. Mowing is the closest I come to working the land and bringing in a crop, even though I mow without bagging, leaving the clippings to return their nitrogen.

The roaring of the engine is an element in the pleasure, silencing everything else. It is a shell of sound keeping out external influence. It's meditative. Going up and down the rows, laying down the pattern, I move into a lawn-mowing trance, my brain waves lengthening.

"The fact is the sweetest dream labor knows" Frost says in

"Mowing," a poem (among other things) about the scything of a meadow. For Frost it's the whisper of the scythe that brings this knowledge, not the droning of a four-cycle engine, but I think I know what he means. Working in a field or a yard, repeating the same motions over and over, our thoughts subside and the plain fact of things emerges. The ground is hard. The trees and flowers have a presence, a density.

Mowing is masculine. Mowing is mechanical. Running the mower is the first mechanical skill I learned—put your foot on the rim here, my Dad taught me, adjust the choke so—and I remember feeling accomplished and grown up as I practiced in our backyard, overlapping the wheels just slightly at each new row, smiling with authority.

Mowing is the first way I earned money, too, first at Mrs. Gray's house next door, then across the street at Mr. Crook's and Mr. McMullen's. I charged $2 per lawn the first summer and kept a little ledger to record my earnings. The next year Dad co-signed a loan so I could buy a new red Toro and go into business in a bigger way. My brother and I were "Anderson Brothers Enterprises Ltd." We put up a sign on the maple tree in the front yard and ran an ad in the *Northtown Shopper*. I was driving by then, an old '49 Chevrolet panel truck, and we'd do hauling and general odd jobs, too. There was nothing better than lifting the mower into the truck on those sweet summer mornings and bombing off to the next job.

But now comes *Fifty Simple Things You Can Do to Save the Environment* and a set of different facts. An acre of lawn needs more than 27,000 gallons of water each week. Americans routinely overwater by 20 to 40 percent. Lawn care increases water use by as much as 30 percent in the summers. Homeowners use up to ten times more toxic chemicals per acre than farmers, up to 5–10 pounds per lawn—a national total of 25–50 million pounds.

And there's no mention here of the amount of gas consumed by lawn mowers across the country, and the pollution that causes. Several friends have decided not to mow their lawns at all this summer. The grass dries up anyway in July and August, they argue, so why bother? Greenness and smoothness are bourgeois affectations.

Mowing the lawn again the other day I was thinking that the 90's are complicating all our usual losses of innocence. It's not just children growing up and discovering their realities are appearances. Grown-ups keep having their rugs pulled out from under them. Taking showers, flushing the toilet, using deodorant, driving to work, eating a hamburger, making a fire in the fireplace—in a time of scarce resources everything we do has implications. Caught up in the grand ecology, no single action can be innocent, unreflective.

"There is something about the human mind that keeps it well within the confines of the parish," E. B. White says. Our habit is to think in "small, conventional terms." But as the world gets smaller and the atmosphere more fragile, the differences between the parochial and nonparochial also shrink. Everything is parochial in the sense that everything concerns us, affects us, just as we affect everything else.

Frost uses the scything of the meadow as a metaphor of death and mortality, echoing Andrew Marvell and the great scythe of time hurrying near. We have always thought metaphorically, always realized that our private lives are implicated in something larger. But anymore that habit of metaphorical thinking needs to be literalized. The mowing of our own particular meadows or lawns not only figures the destruction of the rain forest in Brazil but also has a real effect itself, a literal effect, however small. Blake urged us to see the whole world in a grain of sand. In the delicate balance we find ourselves in, grains of sand might make all the difference.

It's hard not to feel a little haunted—shadowed. The personal and the global are entangled in ways that make it impossible just to live and be happy.

But that ecology can work the other way, giving hope. Interconnectedness means that we have some power. It means that deciding whether or not to mow the lawn and how often is a political act, which is the point of all the recommendations in *Fifty Things*. If all of us composted grass clippings, we could cut landfill congestion by 18 percent. If 100,000 of us avoided over-watering in the summer, 5 million gallons of water would be saved. If even 10 percent of us began using organic pesticides, there would be 2.5 to 5 million fewer pounds of toxic chemicals in the environment every year.

In "The Reactor and the Garden" Wendell Berry reflects on what has the most political force, getting arrested at a nuclear power plant or going home and planting a garden. Both are necessary, he thinks, but it's the garden that might make the most difference. If everyone planted a garden the world would be a different place, not only because of the direct environmental impact—the impact on food distribution, the release of all that extra oxygen into the atmosphere—but because of the calming, centering effect of working in the soil. "Gardening is the finest sort of challenge to intelligence," Berry says, "correcting the cheap energy mind." There's a payoff for our personal ecologies, in other words: the smooth functioning of our own individual systems.

My first garden continues to grow. I put out the peppers too early this month, but the bush beans and rhubarb are doing fine, and wild asparagus keeps coming up at odd places among the rows. The struggle is with the poison oak snaking through the fence, and with thousands of little sow bugs ravaging the slight cucumber V's. Pesticide or no?

Berry's point is that personal health depends on social and environmental health. And vice versa: we can take care of pollution only when we begin taking care of our inner pollutions.

This idea of connection and synergy helps put a book like *Habits of the Heart* in perspective, too. Bella and the other sociologists argue there that American culture is dominated by individualism, that we have no vocabulary for describing commitments to anything larger than ourselves. Our values are private. "We are concerned that this individualism has become cancerous," they say, "that it may be destroying those social integuments moderating its destructive tendencies."

I'm taking that statement personally. After all, it was only when the College of Forestry began harvesting trees in my own backyard that I started thinking at all politically. It's only now that the huge firs have fallen and the heavy equipment come and gone that I've stopped to consider the politics of mowing or the implications of any of my daily habits. My tendencies have always been toward the private and the aesthetic. The forest has always been a place to walk and breathe in the smell of fir, listen to the sound of the stream, not a social challenge, not a responsibility.

Yet the news about the mowing of lawns is heartening because it suggests how movement along one strand always vibrates the whole web. We participate in systems, whether we want to or not. Even walking in the woods has a political effect, if only because it calms us enough to focus on the work there is to be done.

In any event, I am adjusting the level of my mower and I am starting to compost. I'm mowing and watering less. It's not easy. I'm the generation on the cusp, with all these private associations and nostalgias to reverse, and a fatal love of greenness, of the broad green slope. Mowing still centers me. It's that contradictory love of the sound of the engine, so out of keeping with my usual aversion to noise and machinery. It's the smell of the gasoline as much as the smell of cut grass. It's the long hour of the soothing mechanical mindless motion and then coming in for coffee, my grass-stained shoes left on the porch. It's my dad telling me to put my foot on the rim and then yank up on the cord at a slight angle. It's the illusion of pattern, the illusion of authority.

But in the ecology of whatever large systems are at work here, there will be some compensations, I expect. For whatever losses there will be gains—if nothing else an awareness of the web it-self, an awareness of this unavoidable economy.

It's hard to be taken seriously as an environmentalist when you drive a '69 Buick Wildcat. My wife's grandfather gave it to us for free, which is the first reason I could hardly say no, even though it's got a 430 V8 as big as a refrigerator and gets about six miles a gallon, on level ground. When you accelerate you can actually see the needle of the fuel gauge go down. It used to be a luxury car—power steering, brakes, windows, seats, aerial—but it's dinged up in the back now because Grandfather couldn't see very well in reverse, and the front hood, broad as a ship's prow, is rusting out from the rain. It's a dark, Republican blue with black vinyl top and interior, low-slung, limousine-long, faintly sinister.

I'm making do with it as long as I can, since we live out in the country now and can't afford another second car. The odometer reads only 99,357 miles. Fuzzy dice help me brazen out disap-proving stares at stop lights.

I used to be so smug when we lived in town and I could walk everywhere, looking down on the commuters and the two-car families. I've not had to drive before any place we've lived—can measure out my life by the different walks I've taken on the way to work, the neighborhoods I've memorized over the years, house by house, corner by corner. Being able to walk has always been a point of honor with me, until now.

I can maintain a little self-respect because I bike on alternate days, whenever the weather is decent and my legs are holding out. It's fifteen miles round trip, and hilly, just about the upper limit for me, sometimes past it. I love the riding on the good days, get all the expected benefits: the glow of endorphins, the frameless views of fields and trees. I can recover some of my former smugness. I'm self-sufficient again, under my own power, and not dependent on Gulf oil or polluting the air. It's a matter of personal economy, too, of course. Between the bike and the Buick I get about twenty miles a gallon.

Choking on exhaust and catalytic converter fumes in the bike lanes, listening for the doppler effect of the next car, and the next, fighting occasional truck turbulence, I can nurse my hatred of the automobile. I keep thinking of the millions of cars and trucks on the roads today, the millions sold each year, the important percentage of the gross national product devoted to the making and selling of cars, the countless cubic tons of asphalt. I keep thinking of the endless chain of single drivers listening to Madonna and sucking up Big Gulps driving from mall to mall in search of the latest Kmart Blue Light Special. The endless chain, each face blank. I keep wondering whether there's any place of human habitation in the world now out of earshot of the hurrying, elbowing, invasive, Chinese-water-torture whine and roar of arterial traffic, the incessant downshifting and revving up, the constant mental tailgating.

In the science fiction stories some cataclysm or economic collapse has emptied the suburbs. The roads are cracked and full of weeds, the yards growing wild. Things seem pleasantly pastoral to me. When the radioactive dust settles and people start back into their routines again, they look up from planting corn or drawing water and all they hear is the wind in the trees.

Old-timers out where I live now remember when you could stop in the middle of Highland Drive and pick strawberries. The hump between the ruts was grassy and full of dandelions. It's not the wars these people remember when you ask them about the most important changes in their lives, it's the car, the way it's shrunk and hurried up the world, brought all the city people to the forest. People like me.

Being forced to drive the Buick on the alternate and rainy days has been another lesson in how I participate in my culture whether I want to or not. I have to give in more and more. It's tricky. There's a little justification for pride in making do and delaying the gratification of a sleeker, more efficient engine. But not much. I can't hold myself apart anymore. The ironic fact is that to move out and away from the malls and the Muzak I've had to commit myself to the highway.

It's tricky, too, because secretly I love driving the Buick. I love its cabin-cruiser glide, its liquid turns, the weight and buoyancy of the thing. Driving home I sometimes feel for a moment the way I imagine my father felt when he drove home in his Chevies or Dodges, confident behind the wheel, mature in power. It's a grown-up car, a Leave-It-to-Beaver-Hi-Honey-I'm-Home car.

It's Thanksgiving and I'm a boy and we're driving north into the country, brilliant snow all around, and Dad is behind the wheel. Trees and barns move past, the car heater is whirring, Mom and Dad are murmuring and nodding up front. It's after school and raining and my brothers and I pile into the backseat in a heap of wet coats and books. The wipers are going. "Mrs. Robinson" plays on the radio and we pull off onto the street, merge into traffic, and are underway.

I have a recurring dream, even now, of a huge white Ford Fairlane flying down a desert road. It's a square, futuristic car in the dream, straight and simple as a line drawing, and what I experience as I drive it or sit next to my father driving it (it varies) is the simple exhilaration of speed, of the things flying past. It's a dream without plot, just the feeling of exultation, of some deep pleasure in what I know is the infallibility and wonder of the engine beneath the hood, as if all the claims of the car commercials are actually true, as if the car really is magic, can change your life—power without cost. We are weightless. Things keep flying past, the suggestion of a hill, cactus or sage brush. It's evening sometimes and the sun is going down.

Armstrong and Aldrin walked on the moon the year the Buick

was made, an event I remember with absolute clarity. I was at Scout camp then, gathered with the others around a black-and-white television hooked up to the outlet in the commissary, watching the stiff and ghostly figures from under the ponderosa. There was a full moon that week and I remember feeling nothing but joy in the fact of our *being* there, a joy mixed up I'm sure with the joy of being fourteen and surrounded by trees, paddling canoes and diving off docks and cooking over a fire every night. For me the moon shot involved imagination, not technology. I had no sense of the cost of that power, or of the disparity between that technological achievement and the war in Vietnam, the poverty at home. There were no shadows for me then, or contradictions, even though I was growing up in the midst of them.

There's another odd resonance, too, a private though similar irony, since the Buick always reminds me of the tree farm where my wife's grandparents lived for thirty years, harvesting and replanting the hemlock, living off the land. The interior smells of the alder smoke that clung to the carpet and walls of their house, setting off memory, as smells always do. I am digging post holes. I am stretching barbed wire. I am planting seedlings, cutting trail, wheelbarrowing silt from the drained ponds, hauling skyline over muddy hills, yarding alder for firewood, running a chain saw until my knuckles drag and my crotch burns. I am hiking the logged-off hills, beautiful now in leaning alder, corridors of white trunks and shimmering leaves veiling the huge, old-growth stumps, extending off in the distance toward the ITT-Rayoneer land that has spread and spread until it entirely surrounds this tiny 300 acres.

Grandfather is dead of bone cancer now. Nana is stooped over, taking medicine for her heart. The last time they came to the new house they were too tired to appreciate my own line of tall trees, my own little acre and a half of forest. They smiled when I told them we bought the house because it reminded us of theirs, because we wanted to live something of the life they had lived, but they were too spent to do any more.

All these paradoxical associations. All these levels. The Buick is an example of how complex and patchwork and jerry-rigged life is. How you have to compromise and trade off. How you have to enjoy the ironies and messiness. How we are always taking losses. It's somehow representative of the balancing we've all had to do coming of age in the world of Nintendo and MTV that the moonshot foreshadowed and made possible.

It's connected with time, with history. We are living now, my generation, at the moment of transition, right on the edge of one age and the beginning of another. I keep thinking of this as we approach the end of the century. The cataclysm will come, I'm sure, the collapse, for one or more of the obvious reasons. Maybe the change will come slowly or without upheaval, but I'm sure that in any event it will come and that it will be profound. My kids love riding in the Buick, playing with the power windows. Maybe one day when we are gone and the next age has begun they'll be bending over on Highland Drive picking strawberries or having a picnic in the grass, and all around them will be the sound of the wind in the trees. And looking at their own children they'll suddenly think of riding in their father's Buick long ago and of another, more frantic time when the world was mostly asphalt and shopping malls and always in the air was the sound of acceleration, the shifting of gears.

VI

There are days when I still can't keep my perspective, when the ironies start getting to me all over again.

We move to the edge of the forest to get away from other people, for example, and with what seem entirely personal, interior motives. We come to the woods to live our own lives, on our own terms.

But of course all along we're just dots on some statistically significant curve, bits of data too small to recognize the vast cultural trends we were being swept up in. I keep being reminded: everyone is moving to the edge. According to Jack Lessinger, a retired University of Washington economist quoted in the paper the other day, thousands and thousands of people are spilling out of Los Angeles and San Francisco and even Seattle each year and coming to smaller places like Bellingham and Bend and Coeur d'Alene where they can look out their windows and see trees and hills. The suburban way of life is over, Lessinger thinks, especially in the West. People long for slower rhythms, for life close to the land, and freed now by computer link-ups and overnight mail they are deserting in whole percentage points the fouled, frenetic cities and the old neighborhoods with the sidewalks and the lawns for what Lessinger calls "Penturbia," the next "major growth region" to develop since the Second World War. "Penturban" cities are smaller, more compact communities surrounded by forests and fields, hybrids of town and country, with garbage service and browsing deer, cable TV and glimpses of hawks— places just like Corvallis, where I teach and buy my groceries, with outlying housing developments just like mine, partly camouflaged on a forested hill.

So here we all are, thousands and thousands of us, all of us following our bliss, listening to our own inner voices, moving en masse, in droves, to the edge of the forest, looking across our decks at each other every night, hot tub after hot tub as far as the eye can see.

Sometimes it's still disheartening, I have to admit. Sometimes it seems that some Californian comes along and horns in on every Walden.

Sometimes the people don't bother me at all. Sometimes my mood shifts and my angle changes and things appear in some odd, new light.

Not long ago the kids took me down a cutoff trail below our house to a neighborhood I didn't even know existed: house after house and lawn after lawn—maybe a hundred families, with dogs and minivans and lilacs—right in front of my nose all this time. I'd passed the trail a thousand times on my other walks, exploring the roads and the woods, and after a while I'd gotten to feel at home on the mountain we live on, knowledgeable about it. But all this time there was this other place, a place all its own, just a hundred yards or so below and spreading out for blocks, screened by fir and oak.

Our neighborhood is on the middle of the mountain, McDonald Forest climbing above us and extending on the other side for thousands of acres west and north. But there's a terrace below us, too, flattening out to the highway, another neighborhood just as deceptive as ours. All the houses on the mountain are hidden by folds in the land and the distribution of trees so that driving up the highway you may see a roof or two, the suggestion of a meadow, but the main impression is of forested hillside all the way to the clearcut on the top of the mountain. We didn't know our own house was up there until we went to see it the first time.

The sense of stumbling into some dimensional shift was stronger for me because I have such a terrible sense of direction. I am easily lost. I'm the sort of person so pleased and befuddled by the surface of things that I'm continually surprised by new perspective.

It was a soft April day and people were out in their yards attending to roses or mowing the wet grass. Everyone seemed to know each other, and we were often waved at. Cats sunned on decks, birds sang, children rode their Big Wheels. The lots there are an acre and a half or two, like ours above, but grassier and more open, one flowing into the next like a park, smooth green lawn everywhere among the fir.

The houses are modestly upscale Northwest ramblers, roses framing picture windows, daylight basements built into slopes.

Inside we could see the lamps and the bookshelves and the diningroom tables, the cut flowers in vases. I'd come to the woods to escape from town only to discover that the mountain is woven through with driveways and microwaves and barking dogs, a little town in disguise. But it was an oddly reassuring discovery. That day, at least, I didn't mind the company. The lawns and the trees seemed to blend. There were squirrels above the sunning cats, carrying on in the fir—and deer, no doubt, in the thickets, and maybe fish in the little streams, and of course the birds everywhere, the chickadees and towhee, the circling hawk.

Turning the corner on one street—"Fir Glen," I think—we had a straight-on view of the new clearcut the College of Forestry had finished this spring, ugly as sheetrock.

I thought: maybe in the end there'll be more trees in the neighborhoods than in the forests. Maybe in the end urban forestry will be the only kind we have left.

I thought: let's sell off all of the forest right now, in parcels, then commission the best landscape architects to design more neighborhoods that conform to the shape of the land. The timber would be harvested in patches and houses built in the clearings—maybe from those same trees—power and phone lines buried, the remaining trees left in just the right positions to serve as screens and blinds for the far views. The designers would have to be clever about roads, but it could be done. The houses would be soft and New Age inside, with PC's and modems and faxes and cable, but the exteriors would have to fit in with the colors and contours of the forest. All the effort would go not just into building but into blending and merging, camouflaging. I know it's been done in the past: whole army bases built into the ground, invisible from the air. Let's do it here, too, for this war: still the forestry of deception, but for aesthetic effect.

The land would yield a lot of commercial value—what with the harvesting of timber, the real estate transactions, and the construction jobs, not to mention the long-term property taxes— and yet it would still *look* like a forest, more so at a distance.

Walking the roads and waving at the neighbors, you'd at least see stands of fir above the minivans.

"Anyone who observes carefully can hardly find himself twice in the same state," Montaigne says, and this is how I feel about the forest, too. "I give my soul now one face, now another, according to which direction I turn it."

I'm on edge. I vacillate.

Sometimes I wish we didn't live by the forest but back in town where we could have cable and get pizza delivered. We could live right next to the university somewhere, in one of those old houses with basements and porches, and I could walk to school or over to the Beanery. We wouldn't have to make three or four trips into town a day. And mostly I wouldn't feel the responsibility of the forest pressing down on me all the time. I wouldn't feel the responsibility to know about it and learn the names of its parts and understand the ecology of it and its successions and layers of growth and its management and politics. I wouldn't have to worry about what will happen on it next and what I should do about it, what my relationship to it is. I get weary of that kind of thinking, discouraged. I've got enough to do already.

Sometimes I resent the forest just epistemologically, offended that my body and anything external should have that much influence over my mind, my spirit. It just doesn't seem right sometimes—too physical, like a giving in, like a failure to transcend. It feels like I don't have enough freedom. Why should rocks and trees and birds and things matter so much to the spirit, anyway? They know us not. They're just brute extension in space and time. Why should mere physicality—an expanse of hillside, for example, or a section of trees—arouse in me feelings of peace or uplift or well-being? Actually, I get tired of thinking about that epistemological question, too. It doesn't have any answers. I keep spiraling down in my mind and not getting anywhere with it.

Sometimes nature writing gives me a pain, all that rhapsodizing about the mysteries of snails and tree trunks, all that semi-mystical spirit-of-the-land stuff.

Just today as I was walking in the forest I realized that lately I haven't been able to find refuge there because I know it too well, am too comfortable with it. I don't have enough transition into it anymore, since I've moved out here. I just grab my hat and coat and walk out my door, and I know my way, so I'm not preoccupied and diverted by logistics or strangeness. My thoughts flood back. The self-absorption returns. I've contaminated the forest with my presence, my mind. It's time to move on.

Sometimes I wish I owned the whole forest, all 12,000 acres. The first thing I'd do is move my house to the middle, on the Soap Creek side, looking west through the valley to Mary's Peak, where there's the illusion of no houses or farms at all, just forested hills. I'd be surrounded by trees and land, then, and I'd have control over what happened. I'd lock the gates. I'd be protected. I wouldn't have to listen to the traffic noise rising up from the highway, and I wouldn't have to hear my neighbors starting up their Volvo.

In fact, why not imagine a forest even deeper into the Coast Range, far from any towns or people? This one isn't good enough anymore. It makes me mad. It's too close to the highway, there're too many other people around it, there's too much cutting inside it. It's not lush and deep enough anyway—we need to be farther west, where it rains more. I'd like more hemlock, for one thing, and less poison oak while I'm at it, and more old-growth. We need to be farther away from the historic fire patterns, in higher, more protected draws. I want the big trees and the deep recesses, not all this brushy stuff, all these ratty fifty- and seventy-five-year-old stands.

And sometimes, like today, I climb up to the top of the hill behind the house, out of fog and mist and into winter sunshine. I look down on the valley, and it's covered with clouds, like a long box of blankets, but above it where I am is blue sky. The sound of traffic rises up from the highway, but to the east I see the clear blue ridge of the Cascades and all the peaks, the Sisters, Jefferson, Hood. The sun is shining and the wind is blowing. I'm out of breath. Max is nosing around in the underbrush.

Sometimes living on the edge is enough. Sometimes all I'm responsible for is the particular moment. All I own is what I'm seeing right now. Sometimes the only pressure I feel—and this is the right pressure—is the need not to squander this particular sun, this particular forest coming out of mist. Sometimes the only pressure is not to want too much.

VII

In a letter to a friend Keats wrote that the setting sun always "set him to rights" and that "if a Sparrow come before my Window I take part in its existence and pick about that Gravel." That's the kind of man he was, all his biographers say, generous and unselfconscious, capable of entering into the life of other things, becoming something else for a moment, feeling what it felt. There's a famous story about him just naturally starting to hunch his shoulders and paw the air as he described a circus bear, unconsciously taking on the attitudes of a bear, curling his lip, half-snuffling.

I've always lacked that quality of "negative capability," as Keats called it, the ability not only to live with unresolved contradiction but to project myself into others, animals and people and things, imaginatively taking on another life. I'm too self-absorbed, more like Wordsworth, with his "egotistical sublime." The only thoughts I seem to have are my own.

But lately I've been watching the chickadees and nuthatches and dark-eyed juncos and Steller's jays at our bird feeder and on the deck, occasionally a varied thrush, and now and then I have something like the sensation of pecking around for millet and sunflower seeds. If there's anything I might forget myself for, it's birds.

There's much to envy in black-capped chickadees, lightness and buoyancy and instant, thoughtless flight. Watching them from the window, their flashing and quick-cutting, their scalloping back to the maple, I am relieved for a moment of my own heaviness. For a moment I can almost feel that hollowness of bones, that heft of feathers—or the gulfs of morning air and the rush of branches and everywhere the easy purchase, the sure positioning. For a moment I can glimpse the tops of fir like a single surface, a broad sheet of green, smooth and weightless as water.

There's much to envy in the Swainson's thrush, singing of summer in "full-throated ease" like Keats's nightingale. It's winter as I write this, and the Swainson's thrushes are somewhere in Central America, obscured in alien foliage. But they'll be back in the spring, and in the meantime I listen to their beautiful "spiraling" song on an Audubon cassette, reedy and hollow and echoing of deep forest, of sweet dusk in the shadows of trees. It's a song that never fails to lift me out of myself, out to the edge of my own hearing. Listening to it I am always on some far periphery.

The Swainson's thrush belongs to the same family as the European nightingale—both dull brown and shy, despite their songs, hiding in the underbrush—and I imagine the Swainson's when I read Keats's great ode. I think, too, that I understand the longing the nightingale aroused in Keats, though mine is less desperate. I haven't experienced the disease and suffering he did, the financial obstacles, the setbacks and disappointments, and I'm over ten years older than he was when he died. Yet sometimes, too, I long to "fade away" with the Swainson's thrush or the chickadees or the dark-eyed juncos into the "forest dim," hiding in the trees from the "weariness, the fever, and the fret" of life day to day. Like Keats I envy the spontaneity of the birds as they sing, their instinctive and flawless execution, and I wish, too, for their mindlessness, their freedom from anxious, petty thought.

One difference is that I'm not "forlorn" when the birds fly away. Keats's imaginative identification is so intense that the fading of the nightingale's song leaves him sad and confused, lost between idea and fact. (Was it a vision, or a waking dream? . . . Do I wake or sleep?) My imagination, as I say, is never that strong. I just put out more seed, turn to other things, and the next day or the next, looking up from a page or buttoning a shirt, I see the birds gathering again in the trees, the jays sweeping down like caped avengers, and I feel again that small, familiar flush of pleasure.

It's a question of degree, I guess, or maybe of different perspectives on the same thing. Keats wants to become the nightingale, to enter wholly into its otherness and so forget himself. For me the birds are just pleasant surprises, and the pleasure comes because they are separate, not me, uncontaminated by desire.

I think of that early summer morning when I first experienced the riots and festivals of bird song in the forest behind my house, guided by Carol Chambers, the graduate student in Wildlife Management. All this time they had been there, carrying on their separate lives, the hermit warblers and the Wilson's warblers, the flickers and the sapsuckers, the winter wrens and the white-crowned sparrows and the olive-sided flycatchers and the Western tanagers and the crossbills and the purple finches. All this time a thousand lives had been going on in the air and on the ground all around me, with colors and melodies and cycles all their own, each level and species distinct, and I hadn't known. I'd had nothing to do with it. I hadn't planned for it, I wasn't responsible for it, there was nothing I could do about it, and I think the pleasure I felt was a response to the fact of all that intricacy, all that structure and melody and coherence going on completely in excess and to the side of me. I had no instinct to become the birds. They were wonderfully nonhuman, complete without me, and I struggled to sort out their songs not in the hope of somehow harnessing their music but more to appreciate its difference, its independent, spontaneous harmonies.

On another level we can have everything to do with the singing of birds, of course. After all, Carol's dissertation is a study of the effects of the patchcut and shelterwood harvesting. We traipsed from opening to opening, hour after hour, so that she could record the approximate number of birds within a given radius. The damage, so far, isn't clear, although the deep forest species, the thrushes and the warblers, are apparently moving deeper in, leaving the openings to the sparrows.

We should be worried about this, I think, and we should do something about it.

But this afternoon, when the ruby-crowned kinglet was flitting in the top of the oak—or any afternoon, when the juncos scatter, white tails flashing—or in the evenings, when the pileated woodpecker starts rapping on the dead fir, rapping and staring, cocking its mechanical head—whenever I hear or see a bird now, what I think in that instant is: gift. The condition of my pleasure is having nothing to do with its frequency and sources, other than putting out seed and waiting my turn. The birds come and go of their own accord, in their own time, singing their own songs, with or without my bidding. And they appear and disappear, in their color and melody, in their cycles and seasons, so fleetingly, so quickly, that I can't even plan my own seeing. I can watch and I can listen, but I can't make them come. I can't make them stay. They are simply gifts—coincidences—accidents and random glimpses—recurring anomalies—brief spinoffs from vaster and subtler societies of living things flowing and overflowing deeper in the forest, out of sight.

Excursions

The places we have known do not belong only to the

world of space on which we map through our own

convenience. None of them was ever more than a

thin slice, held between the contiguous impressions

that composed our life at that time; the memory of

a particular image is but regret for a particular

moment; and houses, roads, avenues are as fugitive,

alas, as the years.—Proust

It was Fourth of July, and hot, when my brother and I watched the man drown. I was home for a vacation, and Ted and I decided to go bike-riding in Riverside State Park along the banks of the Spokane River just outside the city. It had been one hundred degrees for a week and the ponderosa woods by the river were brittle and dry. Blasts of pine-smelling woods hit us as we rode.

We had gone farther than we intended and were soon at the Bowl and Pitcher, a bend in the river where a large cylindrical rock and a squatter, flatter rock sit side by side among the eddies and rapids. The water is very swift there. We left our bikes in the brown grass and stood on the footbridge, dipping and swaying as we talked, the sound of the rapids swallowing half our words. About two hundred yards downstream I saw a group of five or six people partying on a small strip of beach and then, out of the corner of my eye—I was turning back toward Ted—I glimpsed a man slip into the calmer water along the beach, floating and splashing on his back.

We grew up in Spokane and know the reputation of the river. People drown right at that spot, at the Bowl and Pitcher, every year, and I said something like that to Ted. Gradually, as we talked, I began to realize that the man was in trouble. He had drifted from the calm water by the shore into the swift current and had started to struggle, waving his arms and apparently fighting against the pull of the rapids. His friends on the beach hadn't seemed to notice, or weren't reacting. There may have been one or two people standing there watching. The roaring of the water kept us from hearing the shouts, and the group was far enough away to make faces unreadable. Looking back now the moment seems frozen, a split-second freeze-frame: the beach, the friends standing, the man suddenly in the current, his arms upraised.

We kept reminiscing. The scene in the corner of my eye was at first simply an irritant, hardly impinging on the flow of conversation. On the periphery like that it was hard to interpret exactly what was happening. Even when we stopped talking and started watching the event unfold, we were still uncertain what to make of it. From that distance, and in the enveloping sound of the river, we couldn't separate out the elements of the scene, isolate the details. There was an air of unreality about the man struggling, then panicking, then going under and back up. The rituals of death are cliched to us now: the villainous cowboy clutches his chest and spirals down to the street. I thought for a moment that the drowning man was pretending to drown. His gestures seemed exaggerated for effect. He flailed and gasped and sputtered like a cartoon character.

Someone tried to swim out to help but got caught in the current and had to cling to a rock several yards downstream. A few minutes later a girl in the group on her way up the trail told us that they were trying to get the drowning man to give in to the current and float to the rock, but apparently he couldn't hear them or was too terrified to surrender himself.

Several seconds after we finally accepted the reality of what was happening, the drowning man went under for the last time. It wasn't dramatic. He just didn't come up again. His friend on the rock seemed to be shouting; the rest of the group was by now standing on the edge of the shore, leaning over as far as they could. We weren't sure even then that he hadn't surfaced somewhere else or made it to shore. Only the sudden movement of the girl as she bolted up the bank toward the ranger station signaled his death.

What strikes me now is how little the drowning affected me at the moment. My brother was white-faced, shocked, but somehow the event was already an abstraction for me. Perhaps it

simply hadn't registered. Perhaps it was the literal distance between us and the drowning man that led to the kind of aesthetic
distance I was able to maintain. Perhaps all the stylized images
of death I have seen on television have made me immune to the
real physical fact of someone drowning. Already I was inter-
posing a layer of words and analysis between the event and my
inner life.

And this is partly why I didn't act, why I continued to stand
there on the bridge doing nothing even after I realized the man
was in trouble. There were simple, practical reasons for our pa-
ralysis, as we realized in conversations with friends and family
later. As amateurs we couldn't diagnose the event soon enough
to act. We failed to respond because we lacked experience and
training in emergency situations. As it turned out, the river was
so swift that even when the fire department arrived, it couldn't
immediately rescue the man still stranded on the rock and had
to ask a kayaker to take out the lifeline. It took the rescue squad
over an hour using a helicopter to find the body of the drowned
man, which had been swept a mile downstream. We couldn't
have done anything to help even if we had managed to cover
those two hundred yards or so over rocks and down the bank
before the man went under for the last time.

But we should have started running toward the shore as soon
as we realized something was wrong. At the time we didn't know
that there was nothing to be done, and that shouldn't have mat-
tered. What counts—or would have counted—is our instinctive
willingness to try to help.

Underneath the sense of abstractness I felt, underneath the
simple fear—and there was fear, a fear of drowning myself—is
something far more disturbing for me. My real fear was fear of
embarrassment. I didn't run from the bridge and try to help the
drowning man because I was afraid of making a fool of myself.
In the intensity of that moment the situation resembled in my
mind nothing more than a party I was afraid of crashing or a
group of strangers I was shy of introducing myself to. What if

we sprinted over there in a great show of concern and competence and found that nothing was wrong? My dread was of those strange faces looking up at me as I intruded. As I looked at the scene from a distance, the group seemed closed unto itself, self-sufficient. My instinct was to stay away from it. My sense was that I didn't belong. Even the fact of a man drowning didn't, in the split second of its happening, break through the barrier of exclusivity or privacy that I felt existed between us and establish a sense of our common humanity.

In an odd way I felt that the drowning man had intruded in my life. My brother and I were involved in an intimate conversation. There was a flow to my words that I instinctively wanted to protect as soon as I glimpsed the struggling of the man in the river. Somehow in the first few seconds of his drowning the momentum of that thought was stronger than the momentum of the river sweeping him away.

We stopped at a Circle K on the way back for something to drink. The air-conditioning made us shiver as we walked in sweating from the heat. We heard an ambulance go by as we paid the cashier, who laughed when we told her we had just seen a man drown. I guess the phrase sounded odd to her, like the beginning of a joke, or perhaps she, too, was lost in private thought and instinctively resented being interrupted.

I wonder if the drowning man felt interrupted in the midst of his terror, or embarrassed suddenly to be intruding on the lives of others. I wonder if his thoughts in the act of dying were as mundane and oddly abstracted as mine in the act of watching him.

The first time I fell this year was when I saw my nine-year-old drop off the edge of a log and disappear (I thought) into the Alsea River. I lunged after him, slipped on a rock in the shallows, fell in, jumped up dripping—all this took no more than a second—then leapt over to the island where I had seen John drop. He was fine, it turned out. There was dry land on the other side of the log. I was sopping wet, though, and there was a hematoma on my shin big as a baseball. It was an ugly thing, mottled blue and hard, and it greatly impressed the kids.

The second time I fell this year was at a Jesuit retreat house on the Oregon coast. I had spent a quiet, thoughtful weekend walking in the wind and the alders. It was late at night. I'd been upstairs in the ramshackle house of the sanctuary, drinking wine and reading in my room, and I was coming down to put my glass in the sink. The lights were out. I could hear murmurs from the livingroom. Then my stockinged feet slipped on a step, I felt my body falling at an angle to the stair, I put out my right arm as a pivot, trying to brace myself and turn, like a figure skater (I thought later) making some sort of low, sweeping move on the ice, only my elbow crumpled and the glass went flying. I think I cursed. By the time I regained my feet someone had turned on a light and I could see that the room was full of college students sitting around talking at the end of their first day on retreat. Several came over to help me gather the pieces of glass and see if I was OK before I could mumble my way back upstairs.

The third time I fell this year was in Mt. Angel, Oregon, at the Benedictine abbey. We were there for a week-long conference on family spirituality—a wonderful experience—and we'd been trying to keep track of our four-year-old since Monday. Once again he'd disappeared. Running downstairs from our dorm room after searching the other floors, I saw him playing on the far side of the grounds, in front of the fountain by the abbey church. I ran across the lawn, grabbed him, tucked him under my arm, and started running back, scolding as I went. About halfway I started losing balance. It was a slow fall this time. Even near the end I thought I was going to be able to pull up, but I was too horizontal, Timmy's weight too much. I managed to rotate my torso enough to break his fall, but otherwise it was just a long, awkward-looking dive into the sidewalk, and I drew a crowd doing it. I could see slightly sickened faces out of the corner of my eye as I sprung up with Timmy still in my arms and limped back to the dorm, bleeding from my right knee and left knuckles.

My children fall all the time, as children do, but I think it's remarkable that I've fallen three times in a year myself. I keep wondering about the relationships among these three falls—two having to do with being a father, two (there's overlap) having to do with being on retreat. I'm trying to figure out what this all means.

The symbolism of falling isn't obscure. Fathers often stumble with their sons, people on religious journeys stumble and fall and have to get up, and so we need to have humility, forgiving each other and ourselves.

Another meaning of my falling seems to me literal, not symbolic. It's physical. In my long, slow dive at the abbey, in that moment of being horizontal, I was aware suddenly of how separate my mind and my body are. I was aware of still thinking what I'd been thinking before I started falling—my irritation with my son, and on another level my thoughts about the conference and Fr. Jeremy Driscoll's keynote addresses and the experience I was having of living at the abbey, sharing in the monastic life—these two trains of thought still going on, independently, even while my body was parallel to the ground and I was sticking out my hands to brace myself. It was like the engines still running and the wheels still turning after the train has gone over the cliff.

What I was thinking, too, in the instant I realized I was falling, was that the body has its own life. That it's entirely unconcerned about the acts of the mind.

What I felt getting up, naturally, was embarrassment and chagrin. Most days we walk along forgetting we have physical form, affecting whatever air, thinking whatever thoughts, an abstracted look on our faces. Or we're feeling fine. We're looking good. But then we slip on the banana peel—we're the man in the three-piece suit, with the bowler—and wham, the secret of comedy: the high brought low, the coherent scattered, dignity exposed. It's comic, that is, for the people watching. Forcibly, suddenly, sprawling on the sidewalk, the contents of our briefcase flying into the street, we are reminded of the arms and the legs and the bones we move our minds around with, sore now and bruised.

This is the second stage, the mind not just separate from the body but reduced to it, brought down to its level, and we laugh instinctively—falling or watching a fall—to cover up our uneasiness in the split second we recognize the gap. I've read philosophers who think that this laughter even hides an instinctive cynicism, as if all along we've secretly doubted the attitudes we affect by being human, all the poses we strike; as if deep down we've known there was only the body and the truth of the body. We're just beasts after all, not angels.

Fr. Jeremy wouldn't accept this interpretation, of course. It's God who has the sense of humor, he would say, a profound and loving sense of humor, lowering Himself through the incarnation, taking on our physical, accident-prone form, deliberately taking a fall for our sake. The fall of humanity in the garden leads finally to the entering of God into history in the form of Christ, who stumbles on the way to Calvary under the weight of the cross, then struggles back to His feet and resumes the climb up the hill. Some of the spectators laugh as He hangs on that cross. Only the nails keep Him up, and for a moment that seems grimly funny.

But for Fr. Jeremy there is the rising up on the third day and the ascension into heaven, the reversal of all falls, the completion of the Christian comedy. Jeremy's a slight man with a dark, boyish face, a quiet smile, and he preached to us those five days with great learning and eloquence, his Birkenstocks sometimes teetering on the edge of the dais as he bounced and gestured in his black monk's robes. (Once I saw him almost lose his balance, then catch himself.) What he kept urging us to believe in those morning lectures is that in the incarnation, in the flowing of the trinitarian life from the love of God, the body and the mind can never be separated or reduced to one another. The body has been redeemed.

Sitting there listening, still smarting from the embarrassment of my public fall, my public anger at my son, I was admiring his great and clear conviction, envying it.

      Another falling.

I had been invited to the conference at the abbey to teach a course in the poetry of Gerard Manley Hopkins, a nineteenth-century English Jesuit. There were accountants and pharmacists and mothers and retired engineers in the class, and they were hungry for what literature can do. It was a fine, nurturing experience for me as a teacher.

The third day I was teaching "Hurrahing in Harvest," a poem about looking at the sky and the fields, and I decided to take the class outside to the edge of the abbey grounds, overlooking the whole sweep of the valley. The monastery is built on a hill, with a wide, unobstructed view of the hop and hay and cucumber fields, the irrigation sprinklers shooting out their jets of water, the tiny trucks gliding on the country roads, and the mountains in the far view, their blue and forested sides. It's a view very like the one Hopkins must have had from the Jesuit college at St. Beuno's in the north of Wales, where he did his final three years of theology. The pictures show towered buildings on a hill and the "plotted and pieced" farmlands all around. It was there, in the spring and summer of 1877, right before he was ordained, that he walked and prayed and wrote his handful of great nature poems in a single, ecstatic burst.

We were a month earlier (the poem is dated September 1st, 1877), and we were looking at new-cut hay rather than the "stocks" or sheaves of corn, but it was the underlying movement of the poem I was interested in that day. From the edge there, by the stone wall, you can take in the whole sky at once, it seems, feel the sense of being lifted up and expanded, enlarged, that Hopkins is trying to describe, and I wanted to read the poem aloud with the class looking at that sky and those clouds and how the fields and far mountains seem to blend into them, all blend and merge and swell, sweep up:

Summer ends now; now, barbarous in beauty, the stocks arise
Around; up above, what wind-walks! what lovely behavior
Of silk-sack clouds! has wilder, wilful-wavier

Meal-drift moulded ever and melted across skies?
I walk, I lift up, I lift up heart, eyes,
Down all that glory in the heavens to glean our Savior.

Our instinctive response at the sight of the open sky and fields is like the lifting up of the heart in the mass ("Lift up your hearts!" "We lift them up to the Lord!"). It's a sudden welling up and widening out of feeling, and for Hopkins, instantaneously, a revealing of the presence of God, the sky—like the whole world—forever sacramental. He "gleans" the savior in that sacrament, harvests the sky as he would the fields, winnowing and delighting.

At the end of the poem is a fall—a rapturous one this time, the result of joy. Reading it aloud I found myself for a moment very moved, teetering like Fr. Jeremy on the edge of the dais:

These things, these things were here and but the beholder
Wanting; which two when they once meet,

The heart rears wings bold and bolder
And hurls for him, O half hurls earth for him off under his
    feet.

At the sight of Christ revealed in the sky and fields, the heart grows wings and seems almost to jump out of our chests, to burst out of our bodies—when we as "beholders" "meet" the world, see it with imagination and love, our feelings "hurl" themselves outward. The earth seems to move from the force of that longing, but more literally, I think, we are the ones who move, who are knocked on our keesters. The effect of the bold heart hurling is that our legs fly out from under us and we fall to the ground, like St. Paul knocked off his horse on the road to Damascus, but joyfully, willingly, not blinded from above but expanding from within in the (futile) effort to fly, to lift up.

It would look funny to someone watching, slapstick—a man flinging himself to the ground, just throwing himself backward, feet above his head, like a clown.

There are further possibilities, too, in the first two falls of the year. The fall at the Jesuit retreat house was just a silly anomaly, for example, a stockinged slip in the dark, but the lunging and the hematoma at the Alsea River were the result of a father's unconscious protectiveness, a sudden, physical leaping out. As an instinct it was purer than the complicated layers of feeling I have for my son consciously, when I have time to think. It was a clean hurling forward.

It was autumn, and we had come to see the falls at the river, which roar down from the rocks and fall home to the quiet pools like the Scottish "burn" Hopkins celebrates in "Inversnaid." ("This darksome burn, horseback brown, / His rollrock highroad roaring down, / In coop and in comb the fleece of his foam / Flutes and low to the lake falls home.") Leaves were falling from the maple and alder, yellow and red against the hills of fir.

It was early spring when I went to the retreat house on the coast, raw and windy but mild. The house sits on the end of a peninsula surrounded by Nestucca Bay, the ocean just beyond, on the other side of the Nestucca Spit. In March it is a wet, gray place. The Jesuits logged the old-growth fir in the fifties, and the forest now is mostly alder, mossy and damp in the coastal rains. Fern and lichen grow from nurse logs. Fallen branches break like graham crackers under your boots. Everything is rich and de-cayed and tangled.

If God comes to a place like this He comes not in a sweep-ing, lifting motion but "dappled" and "pied," through the white trunks and the crisscrossing branches, through the leopard slugs, the wet skeletons of leaves plastered on the trails, the thousand thousand mottled things.

One day on a walk around the peninsula (now that I think about it), the wind blowing and clacking in the alder, I followed the impression of a logging road off the main path, then a deer trail, then found myself against a low embankment. I tried to scramble up, pulling on sword fern and salmonberry, but the roots came out and I fell backwards, back onto the soft, forgiv-ing ground, the wet moss. Later when I took off my coat I saw that it was covered with moss and clay, like a coat of earth, a coat made of the forest floor.

The root of "humility" is "humus." Earth. To be humbled is to be brought down to earth. To have your nose rubbed in it, maybe. To be covered with it. It is to fall.

A monk was once asked, "What do you do there in the mon-astery?" He replied: "We fall and get up, fall and get up, fall and get up again."

It was hard to leave the monastery, to come back down from the hill and enter into the world again after a week of living in community with other families and sharing in the life of the monks. Everything seemed sad and ugly ("smeared, bleared with toil").

One image I carried away with me is of the little graveyard where the monks are buried, the simple stone crosses lined up in rows. Around them is a little grove of fir and maple, and rising right behind, like a B-movie spaceship, the abbey's water tower, a large sheet metal drum on a three-story tripod. We had been hearing an odd sound coming from that direction at night, a high-pitched moaning, and we decided it had something to do with the water tower, maybe with water running through pipes. It's ghosts, Fr. Jeremy joked when I asked him about it, but in fact it's the sound of the wind blowing across the struts and supporting wires around the drum, which then picks up the motion and begins vibrating, like a glass of water hit with a spoon. The wind blows more on the hill than in the valley, and it freshens at night, so the music of the tower was a constant evening sound, what we heard each night when we went to sleep, more vivid in memory than the bells that summon the monks to prayer.

It's a wavering sound, as I hear it now in my mind, more like a musical saw, bent across the lap and braced, then stroked with a bow, vibrating unsteadily, mournful, hokey.

New Jersey is the objective correlative of all I dread. It's the future: glass and steel and asphalt, oil refineries and bombed-out factories giving way to office parks and townhouse developments, four-lane freeways every few blocks, the cumulative effect as Byzantine as *Blade-Runner*. Sequestered in a Marriott Hotel with two hundred other advanced placement exam readers I lived pretty much like the rest of New Jersey in stifling June, as far as I can tell. Everyone there seems cocooned in machines and artificial environments, going from house to air-conditioned car to shopping mall or office complex. We might as well have been on the moon.

I know I'm being unfair. There are lovely parts of New Jersey, and the things that repelled me about it are all over the country, even in Oregon. It's a question of degree. Every shopping mall is a bit of New Jersey. Every freeway. Most of what's on TV. The little piece of New Jersey where I spent six days reading advanced placement exams for SAT is simply one spot where the global project is a little more advanced.

Two images.

Sitting in a refrigerated, windowless room around tables covered with white tablecloths, eight hours a day for a week, thirty of us, high school and college teachers, silently reading advanced placement booklets, flipping pages every twenty seconds, scribbling down scores, putting one exam on the growing pile and picking up another, one smooth motion, silence except for the hum of the air conditioner. Periodically we are normed, scoring essays in common and recommitting ourselves to narrow point spreads. The words and sentences blur, swim. The exams pile up. We emerge at break time blinking.

I am at a sushi bar in New Brunswick eating dinner with an old friend. The tables and counters are made of a black glass with silver chromed trim. The lighting is harsh but indirect, concealed in narrow panels. The aisles are dark. Across from us a family sits on stools, lingering over their dishes, a dark-haired woman in a white T-shirt, smoking, her arms thin and pale, her small son sipping what seems to be a daiquiri, the father with his back to me. It's air-conditioned cold. A Japanese man in a white uniform is leaning over their counter in the harsh light, smiling. No one in the family speaks.

But there were surprises on the trip, too. There were anomalies, outer and inner.

Canada geese live at the hotel pond, for example, a flock of about twenty or thirty, and when they took flight at my approach there was that lovely whirring sound of their large wings. A surprise. There between the freeway and an office park a flock of geese, honking and pecking at the grass, defecating on the footbridges.

Or beneath the trees behind my friend's house in New Brunswick a dozen raccoons congregate each night to eat dog chow in the shadows. All around them is the sound of traffic and the glare of streetlights, but there beneath an oak tree: the glittering eyes and the suggestion of masks, the little, human hands worrying the nuggets.

There was always the sky, too, the familiar sky with all its wonderful weather, the Eastern clouds swelling with moisture, the sun streaking through the pollution—a prospect of wideness and openness above the hotel canopies, outside its pneumatic doors. When the future comes and the whole world is like the Somerset Marriott, maybe we will still be able to take comfort in the prospect of the sky and remember the feeling of openness.

I took comfort in the human things I saw. Even in New Brunswick people have kitchens and coffee cups and coupons taped to the refrigerator door. Even in unimaginable New Jersey people read the paper and do the dishes and make the bed before going off to work, and some important dignity attaches itself to all that, to the repetition of those acts over time. This is an obvious truth, the kind that people travel to rediscover. All places are good places—sites for routines and repetitions, the accumulation of our human habits.

Location is the great mystery. This house or this field or this sidewalk in New Jersey has breadth and depth and width at just this time and place. It is nowhere else. Every kitchen or tree represents some miracle of being present. Glimpsed from a bus window or through a hallway at an old friend's house these little corners of light and shadow have a kind of endearing density. Reality is always familiar.

It wasn't long before I was even enjoying the enclosedness and the air-conditioning and the attention to our comforts, everything padded and hushed and sealed—enjoying it despite myself. Meals came regularly, we slept on schedule, the readings proceeded apace. Something in me righted itself. Blood pumped to my arms and legs, food passed through its accustomed tracts, air moved in and out of my lungs, and before long I was over the vagueness of travel and seeing the world, even that world, with my old eyes. I recognized myself. The spring came back in my step.

There I was assuming my misery, reading New Jersey as the embodiment of cultural decline, and then padding down the hall to my room the third day: a flicker of contentment. There was work to be done. There was the rhythm of breakfast and lunch and dinner, conversation with friends, my evening walk. There was all the old anticipation of things.

We must keep chaining ourselves to trees and putting sugar in the bulldozer gas tanks, almost anything to avoid turning more of Oregon into New Jersey. We must resist. But personally, internally, some inner muscle relaxed, some peace flowed into me that moment walking down the hall of the Marriott when I realized that even there I could be happy. Driving the malls on errands some days I feel a bile rising in my throat. I've come to worry about misanthropy, worry that I've left myself no place for forgiveness, since the asphalt and the video stores are everywhere. The epiphany in New Jersey released some of that tension, taught me in some faint way what a soldier or prisoner learns. Something human always survives, looks out from our bodies at the world, distinct from it.

I've always anticipated apocalypse. Maybe it's all the end-of-the-world science fiction books and movies I knew as a kid, or growing up with the nuclear threat, but I've always had deep in my mind some vague image of hiding with my family in a bunker somewhere, waiting out the catastrophe. But in New Jersey I experienced the moral of those doomsday stories. Life is always possible. A flower grows here and there. A tree leafs out. Other people are found in the rubble. Soon a village is going up and rules are being reinvented. Someone finds books. Looking into each other's eyes months after the first explosions, the actors see some glint of humanness return. A man and a woman join hands, smile. The point of all those old movies of Armageddon is not just the inhumanness of our destruction but "the dearest freshness deep down things."

Maybe it's true. Maybe when the future comes and the whole world is like the Marriott, some few of us will remain to join hands and walk off across the parking lots toward the broad and familiar sky.

It's easy to critique the 5,200,000-square-foot West Edmonton Mall, the world's largest. It's 120 acres of what's wrong with contemporary life, all the cheapness and glitz and consumerism of malls everywhere magnified and made even more garish: 210 fashion stores, 55 shoe stores, 35 jewelry stores, 110 restaurants, 2 car dealerships—and an NHL-size hockey rink, a scaled-down version of Pebble Beach golf course, a life-size replica of Columbus's *Santa Maria* floating on a four-acre lake, a separate seven-acre water park complete with palm trees and wave machine, a 560,000-square-foot amusement park, 30 aquariums, over 250 kinds of exotic birds, and a 360-room luxury hotel. The West Edmonton Mall is the culture concentrated and writ large, 48 city blocks of Ninja turtles and VCR's glimmering beneath vaulted glass ceilings, all of it continuous and enclosed, so vast and sprawling it requires five postal codes of its own.

Yet wandering around it one afternoon the last day of a trip to Edmonton I found myself neither awed nor disgusted. Something about the place rang true. It only makes sense to seal yourself off from the oppressive, stagnant cold of the Canadian plains. Three days of ten below and the foul, frozen air and the dull, featureless expanse and the rushing from one enclosed tiny, stale, overheated room to another were enough to make me crave greenness and space. The mall is a biosphere, a substitution for the drabness and ugliness of the actual world, and in it the designers have replicated all the things that Edmonton in the winter lacks—good things, it seems to me, essential things: trees and birds and water and opportunities for unobstructed movement.

In Oregon only porous boundaries separate me from the outer world. Single-pane sliding glass doors and picture windows hold back the mild winter air, and the wind rushes through the cracks. I am always in communication with the weather, can always see or feel the air, the gray light pouring through every room, the sight of rain and the sound of it. In Edmonton the windows are smaller, of course, to keep out the cold, and the insulation much better. There are no gaps in any room or house or building, and the air inside the sealed enclosures circulates and recirculates, the sound of blowers in every background. To get to the outside you have to put on your sweater and your coat and your scarf and your wool hat and your gloves and walk stiffly to the thick doors and haul them open and when you do get outside the shock of the ten-below cold throws you back onto yourself. You huddle deeper in your coat and scarf, hunching down your head, becoming your own enclosure.

The mall is enclosed on such a grand scale it creates an impression of expansiveness. Walking in you feel lightened—freed of coats, able to walk its 1.609-kilometer length in seventy-degree, ersatz spring. The two-story-high vaulted ceilings have something of the effect of a Gothic cathedral, though secularized, flying buttresses replaced by plexiglass honeycomb, grid after grid of industrial latticework as far as the eye can see. A pleasant winter light filters down on the earring shops and McDonald's and the caged albino pheasants, oddly natural, foregrounding objects like late afternoon sun. There are shadows. Potted palms and Japanese maple line the walkways, motionless in the still air. Water gurgles from pools and fountains at every interval.

It's poststructuralism triumphant, I suddenly realized, fact and value completely divorced—value entirely artificial and imposed, the cold, barren real world entirely sealed off, merely the foundation for constructed realities.

Looking through the glass walls of the huge "World Water-park"—the size of five football fields, the brochure says, containing 2,690,000 gallons of water—I felt a quick pang for the real Pacific. There it was in impressive miniature, bathers splashing on a sloped concrete beach, others, farther out, bracing and bobbing on the artificial waves—computer controlled, ten minutes on, ten minutes off, reaching heights of six feet. The enclosure for the Waterpark is higher than the rest of the mall, sixteen stories, and twenty-two slides snake and twist from quite high up, children sluicing and screaming from almost 85 feet at almost forty miles an hour. Closer to us, by the windows, grown-ups lounged in swimming suits around palm-fringed Jacuzzis, sipping Hawaiian drinks in a constantly maintained eighty-six degrees.

At Europa Boulevard people ate crepes and drank white wine at sidewalk cafes, gabled French building fronts ringing off the courtyard. "Take a stroll and discover the bright and airy out-door feeling created by a magnificent glass dome," the brochure proclaims, and I have to admit to the impression of airiness—and to the telling effect of the scrolled wrought iron, the Parisian fountain, the murmur of voices. No doubt there was simple ply-wood underneath the high-tech paint, and crooked modeling on closer inspection, or cracked plastic, but I was content to keep my distance and so preserve the illusion.

For me the mall was an experience of atmosphere, not con-sumption. I spent the whole afternoon without going into a single store, buying a single sweater or CD. It was hard to pay attention to shopping itself, although I'm sure people get used to the scale and excess before long. The Iranian immigrant brothers who built the place no doubt hoped to make a lot of money, and no doubt they have, but I can't help thinking there were other motives, too, aesthetic motives, the material and props of malls swept up in a larger enterprise. Surely sufficient merchandise could be moved without the aid of an 85-foot-high water slide or the world's largest triple-loop rollercoaster (fourteen stories high, with 4,180 feet of track).

All my critiques of malls remain in force, all my resistance. I have no desire to return to West Edmonton. And yet whenever I recall that afternoon I feel again that slight hollowness, that emptiness, you always feel remembering a moment of pleasure.

I keep thinking of the dolphin show I saw at the end of the day, four Atlantic bottlenose dolphins, each approximately four-hundred pounds, powering around a large transparent tank, doing all the synchronized flips and hoop jumping that dolphins do, gulping smelt and herring between tricks. The Plexiglas walls of the tank sit a little higher than the stage, so you can see the dolphins underwater as they streak past the edges, looking out at the crowds with those fixed eyes and fixed, dolphin smiles. It took two or three laps of the tank to get up enough speed for the high jumps. You'd get a glimpse of those marvelous bodies and then from somewhere deeper in the tank they'd shoot up, twenty, thirty feet into the pure light under the central glass dome, shining in the still mall air. The crowd would do its gasps and then the water would splash over the sides as the dolphins hit the surface again, dousing the delighted children.

It's terrible to keep dolphins in a place like that. I can't condone it. And yet in the moment I was entranced. I keep thinking of their white underbellies as they tread water in front of us, the skin so seamless and tight and shimmering it looked like fiberglass. I keep thinking of those smiles, and of the sleek bodies suspended for that second, arcing through the hoop, and the spectators gathered in the little amphitheater by the tank and all around on the levels above, tier after tier lining that huge atrium, holding their breath, ready to applaud, plastic sacks leaned against the railings, tier after tier—those marvelous dolphins, arcing in the still winter light beneath the dome.

Not long after I got married and moved away Mom and Dad sold the house at 5917 N. Lincoln where I grew up and moved to Contempo Spokane, a "manufactured housing" development north of town. Contempo is one of those pristine, regulated, de facto retirement communities, all manicured lawns and pebbled rock gardens and polished vinyl siding—straight-edged, clean, lacking trees.

The single-wide Marlette my parents bought combines the plushness and shininess of a new car with the snugness of a house. Everything in it is new: the sculptured carpet and the wood veneer, the self-cleaning oven and the fiberglass showers, the double-paned windows, the built-in stereo. Mom's pewter clowns and porcelain roses fit in nicely, and she bought new furniture for the move, floral couches and gold floor lamps. It's like the house I grew up in in some ways, has all the same components of "home"—chairs and beds and dishes—but sanitized, translated into some contemporary language of automation and self-containment. Walking in you have to admire the way it all coheres aesthetically, its effect of comfort and efficiency and brand-newness—the ingenious use of space, like the living quarters of a ship, and at the same time softness, domesticity. Glass vases reflect off polished walls. The aluminum sink sparkles. Feet sink in carpeting.

Mom has always been a fanatic housekeeper, obsessed with neatness, and I think that was part of the appeal for her. There are no damp cellars or cobwebby attics in a mobile home, no gaps or ghosts. All surfaces are clean.

The whole community is walled off from the rest of the town, with its own security guard and speed bumps, its own post office. It sits in an open field, fenced-in. There are rules about keeping up yards and the design of awnings. Plastic sunflowers and ceramic Hobbits sit in every front flower bed, or imitation deer beside small shrubs. In the corner of the court is a separate corral for the RV's and campers of the residents, many of whom own mobile homes some place in Arizona, too, and travel back and forth seasonally, going from one mobile to the RV and then to the other mobile, the decor—the plushness—never changing.

Contempo is like a small town. When strangers drive through the main gate their progress is followed. The old-timers go down to the main office the same time each day to pick up the mail and shoot the breeze. People are always talking across back fences or walking the loop—silver-haired men and women in sweats and white walking shoes—or riding three-wheelers. This visit I went to a cocktail party with my parents on Betty and Jim's patio. Betty is a slim, pretty woman my mother's age in sleeveless blouse and shorts, and for a moment as we talked there was something almost girlish about her. She drank her highball and told dirty jokes and talked about her grown children scattered across the state driving trucks or working as receptionists. Her husband is retired from the telephone company and didn't say much. Later he showed me his collection of Charlie Russell prints. It was August, in the cool of the late afternoon. The air was dry. Next door someone was watering.

There's a whole subculture here, defined not just by its flat roofs and waffled awnings or its patio furniture and barbeques but by age and class. Most of the people in the court are retired civil servants, like my father, or teachers or small businesspeople, comfortable without being well-off, and they've come to these communities of vinyl siding and plastic sunflowers to live a version of the good life. They've come for peace and maybe the broad night sky of stars, unobstructed by trees. They've come because their families are raised now and they can no longer afford, emotionally or financially, all the trappings of "home" as most of my generation remember it.

That's the odd, disquieting thing. It's my generation with the nostalgia for tree-lined streets and porches and houses with basements. It's my parents' generation that seems to be embracing the future of manufactured housing and ceramic Hobbits, divorcing itself quite deliberately from the past.

Our last day I persuaded Mom to get out the box with the old photographs. She's never arranged them in albums. Sometimes it's difficult to get her or Dad to talk about their past. But I insisted, and the box came out and Dad started sorting through the pictures, reminiscing. It hit me then what this is all about. In one picture Dad is a tow-headed boy the age of my youngest son. He's standing in the snow in threadbare clothes and behind him is the suggestion of a hill, some trees, and a shack where he lived with my uncle and my grandparents. Grandpa was an electrician and during the Depression had trouble finding work, poaching deer and panning for gold to get by. The family moved from one rickety rented house to another. There was another photograph of Dad and my uncle and my grandmother on a porch and in front an old panel truck with "Clark's Electric" handpainted on the side. The shock for me in those pictures was the starkness and poverty of my dad's youth, how little time—it suddenly seemed—separated my life of comfortable middle-classness from the scraping by and the cold and the almost hunger my dad endured as a boy growing up in northeastern Washington, in Ione and Colville, as my grandfather scrambled for work.

My mother's story is a story of an alcoholic father and a denying mother, and coming home from school to scrub and clean a succession of dingy houses while her mother worked to support the family. There was the deep wounding of her father's, my grandfather's, gradual breakdown, the vow to live a very different life, to have a home of her own where the dust never gathered in the corners and there was always food in the refrigerator. There was the vow not to owe money, not to scrimp and save and can beans and put up preserves. There was deep anger.

The central image in Mom's history is of a house fire when she was younger than my daughter is now. Some wires apparently shorted in the attic and the old wooden house went up fast. Mom and her brother had to be thrown out the second-story window. Aunt Lettie died rescuing her son through flames and hot steam as Grandpa Ted sprayed the ceiling with a hose. The next day the blackened remains were on the front page of the *Spokane Chronicle*, everything destroyed.

My parents don't want to remember the past. They don't want foundations. They've lived their lives for the future, which then moved away, and now they want to forget the blackened remains. The mobile home—the manufactured home—represents their desire: for newness and snugness that is not rooted, that is not permanent, that is cut off from the ground and the trees and anything natural because the natural recalls the flames or the snow or the threadbare clothes. Every time I visit Contempo I get the sense of a whole generation trying to deny foundation, a generation that grew up in the Depression and fought to give their children the comforts they never had and now, the children gone, want nothing more than clean, white walking shoes and the open asphalt circle of the court, complete with speed bumps. They want lawn chairs and highballs and an unobstructed view of the sky in August, and they want their children to visit every summer but not stay—there's no guest room now, no room for the grandchildren—because the struggle and pain of that is over now, too. They want some measure of freedom.

When I was talking with Betty and Jim that day on their patio, and she suddenly seemed like my contemporary and not my parent, I wanted to put my arms around her and say, it's OK. When Dad was going through the snapshots and there was suddenly that picture of him sledding, his body a blur against the line of trees, I wanted to tell him that I understood—that I'm a parent, too, and that I understand a little better now about struggling to escape and build.

It's finally more than their individual histories that my parents are trying to escape. It's history itself. It's time. Most of the pictures we looked at were of my brothers and me as children: in cowboy hats and holsters, riding two-wheelers, visiting Grandma Gottwig—school pictures of the three of us in striped T-shirts and crew cuts, or Christmas pictures with parkas and hoods, throwing a ball or chasing Taco, the dog that ran away on the Fourth of July. What surprises my parents now, I think, what surprises me, is the scandal of children growing up and moving away, growing up so fast there isn't even time to sort the snapshots. I could hear the hurt in my mother's voice as she finally joined our reminiscing. After all those years polishing the coffee table and scrubbing the kitchen and sending us off to school every morning, all those years trying to deny her childhood, she finds herself having to deny our childhood, too. She won't put those pictures in order because she's been betrayed—because there is no order, only the shocking progress of the years, the growing apart. Mom and Dad didn't abandon us, we abandoned them, as we had to, and now in this new present they don't want a basement full of empty bedrooms to remind them of what I'm experiencing now every time one of my three children climbs the stairs asking for breakfast or looking for a lost shoe.

I understand a little better now, Mom. It's the old catastrophe, the one we couldn't believe until it happened to us. As you sit at your formica-topped kitchen table and listen to the windchimes on the carport, or watch the light coming in through the prism, what you're thinking about, I know, is the hurt and mystery of time. We are contemporaries in that, Mom. In that at least we are alike and there is no gulf of years.

IV

Scott and I are splashing up Portland Creek in hip-waders looking at logs the Forest Service has cabled to boulders all up and down the stream, as far as the eye can see. Old-growth forest stretches away on either side of us, part of a Spotted Owl Habitat Conservation Area set aside in the Willamette National Forest outside of Eugene. The water is low, exposing the plaster-like surface of the streambed, the scarred logs placed at carefully determined angles every few feet, some pointing upstream, some perpendicular. Cables bristle from their ends. It looks like some odd playground equipment, I say, laughing—zoo habitat. Behind us a dozen men in red hard hats are standing in the stream drilling holes in boulders to anchor the log cables, the sound of the large drills filling the air. A couple of Forest Service supervisors stand around on the bank, scribbling on clipboards. What impresses me at first is both the incongruity of the sight—the ugly lashings beneath the alder and fir—and the scale of it, the sheer numbers of logs angled and tagged.

Earlier in the summer huge excavators and rubber-tired skidders had powered up and down the stream, bringing in logs from other parts of the forest. Scott makes big, swooping gestures as he explains this, shaking his head and smiling. Boulders had been dynamited from a distant quarry and trucked in, too. Equipment and crews of men were jumbled up everywhere, the creek-bed like a big-city construction site, full of shouting and exhaust fumes.

Decades ago and farther upstream, Scott explains, loggers finished up a large harvesting operation by removing all the logs and woody debris from the water—standard practice for the timber industry then. The idea was to leave things clean. But scientists have discovered now that woody debris creates pools and "riffles" and "glides" that fish and other wildlife need to keep from being swept away. Wood creates "resistance," adds "roughness." It builds "structure." Sill logs and other kinds of logs dam up silt and gravel, too, which provide necessary spawning habitat. And the sill logs trap leaves and other organic matter which feed the bugs that feed the fish.

The irony on Portland Creek is that it's against the law to cut trees in a Spotted Owl Habitat Conservation Zone, even for stream restoration. Logs have to be hauled in like steel girders precisely because the area has been set aside as "natural," to be left alone.

They have to be cabled to boulders, artificially fixed, to keep them from being washed away in the spring runoff: random debris replicated only through precise measurement, deliberate placement.

It was September as we walked up the stream, early leaves

swirling over the structures Scott helped engineer. At the bottom of one deep pool—a "P 5" Scott labeled it—alder leaves shone like pennies. Scott is a "co-op" student with the Forest Service, studying for a Masters in Fisheries and Wildlife at Oregon State, a good-looking man in his early thirties, quiet and considered. He spent the summer working as a designated "stream designer" out of the Lowell Ranger Station, first counting the fish in the stream (shocking pools with an electro-shocker and counting the fish that floated up), then directing the "installation" of the logs according to a complex computer-generated map of the stream, then tagging and marking each log, carefully recording its position and intended purpose. At intervals he'll return to see if each installation has met its "objectives," though eventually, if the project works, many of the logs will be buried by the silt and gravel they've built up.

The conceptual artist Cristo ringed a Florida island with sheets of pink fabric and later stretched a cloth fence twenty miles long across the fields of Northern California. This fall he directed the simultaneous opening of thousands of giant yellow umbrellas first in Japan and then, the next day, in a California mountain pass. As I unstrapped my hipwaders by the side of the jeep it seemed to me for a moment that the cabling of the logs and the designing of the stream were no less a kind of "environmental art," improbable, even absurd—and wonderful, too. It's hard not to admire something so apparently crazy.

I realize that the government's motives are not aesthetic or disinterested (as Cristo's aren't either, since he makes a small fortune marketing swatches of his fabrics and videos of himself in the act of creating). The designing of the stream is part of a "Salmon Enhancement Program" driven by the pressures of game fishermen to have more fish to catch. That's what principally drives most everything the Bureau of Fisheries and Wildlife does, as I realized earlier that day, at the Oakridge Hatchery, watching the spring chinook "spawn." Five men in rubber overalls stood to their waists in a fenced-in holding pond, the water alive with chinook, seething. Two more men and a young woman with a clipboard waited on the dock, the end of the assembly line. Stage one: bellies felt to see if the hens are green or ripe. Stage two: green hens tossed over the fence into a small canal (dozens of bulky tails and heads bobbing off into the distance, like a herd of fording cattle). Stage three: ripe hens whacked in the head with a wooden club (sometimes two or three times for good measure) then bucked up onto the dock and into a slot on a row of plastic trays where the tails are clipped with garden shears to drain away the blood. Stage four: drained hens whipped around from the trays and, almost in the same movement—held by the gills—slit open from tail to gullet to release cascades of bright orange eggs, hundreds in each fish, pouring out into five-gallon buckets. Stage five: the dead chinook thrown onto a growing pile of carcasses in large, plastic-lined wooden boxes. The ripe bucks go through the same cycle, only in the last stage their sperm is milked into the buckets of eggs, squirting out when the flanks are squeezed like milk from a teat.

And it happens over and over again, hundreds of times a day, thousands of times a week during the fall season—the thrashing of the fish, the whacking of the clubs, the slippery-suited men, the shouting and laughing, the bending to the work, the growing rhythm of tossing, whacking, bleeding, slitting, the glassy eyes of the fish, the thump of bodies in the carcass box. The eggs and sperm mixed up in the five-gallon buckets are later put in oxygen-enriched trays where the fry emerge, which become the larger fish in the larger holding ponds, then the larger fish, and so on, until it's time to be trucked or flown and released into whatever river or lake. The same assembly line produces the fish—trout and salmon—for every major water system in Oregon, and has for two generations, ever since the building of the major dams. That's the point. The many dams on the rivers disrupt the spawning cycle and so that cycle must be artificially re-created, season after season, or there will be no fish in the rivers above the dams, no fishermen paying fees, spending money for fancy reels and potato chips and Budweiser and four-wheel drives, voting for their favorite congressperson.

I wasn't disgusted by the spawning so much as surprised. I hadn't given fish much thought. My images of salmon in rivers were images, just instinctively, of the natural and spontaneous, the nonhuman. But here, too, more is happening than meets the eye. Beneath the surface is complexity, and irony, and contradiction, teams of specialists, special interest groups, government bureaucracy, profit motive, contested history.

One of the contradictions is that the conventional spawning and hatchery program is at odds with the new stream designing and salmon enhancement effort. Salmon can't get past the dams and upstream to take advantage of their new habitat even if the installation of logs at their special angles actually works.

Another issue, of course, is that hydroelectric power is some of the cheapest and cleanest we can get, and we need clean power. The question is always how to balance these things out.

But what I keep thinking about, what seems most important to me as I reflect on that day, is how different the two operations felt, how different their effects were on me, as an outsider getting the tour. Both the designing of the stream and the spawning of the chinook are unnatural, human activities, involving plans and jargon and at some points heavy equipment. But I can't help thinking of the designing of the stream as a better thing to try to do. Scott started me out in the morning at the Dexter Dam just outside of Lowell. There were the usual sheer concrete expanses and the chain link fencing and the power lines everywhere and the "no trespassing" and "high voltage" signs, everything large scale, obviously official. A beautiful lake backed up from the dam, it's true, and the light off it was clear and the air sweet, but my impression was more of passing through government gates and walking on government gravel and looking up at big, impassive, concrete structures. More green chinook waited in holding tanks strung off and labeled across the compound, large fish, beat-up and blemished, now and then slapping hard against the concrete sides of the tank, trying to get out, bludgeoning themselves, weakening.

The dam is huge, with far-reaching effects. The sill logs up on Portland Creek, the boulders and the anchoring cables, are small finally, their effects local. Each log is a small dam itself, the water shooting off it into tiny pools, pouring over silt in shallows. Dexter Dam is obvious. It will always show. No effort has been made to disguise it and it will never be buried, short of earthquake or structural catastrophe. But the logs and the cables up on Portland Creek will disappear in a few years, evidence of the human hidden by gravel and cobble and the flow of deepening water—in that sense finally not like Cristo's conceptual art at all, not for show, for display. Dexter Dam is a grand simplification of a system, single and whole, its logic easy to understand. The attempted replication of the stream is exceedingly complex, almost impossible to map, each piece in the system, each force in the dynamic, delicately balanced against the others.

The spawning of the salmon is a simplification, too, and it serves the end of efficiency. It's a mass-production process, generating huge inventory. Whatever else you might say about the designing of the stream, boondoggle or not, it works against the cause of efficiency, and it's for that I want to praise it. Reinventing glides and riffles, building in overhangs and hollows and backups at fifty-yard intervals, is not the most cost-effective way to generate units of fish (the hatchery is). The ratio of salmon-to-dollar is so low there must be other values at stake.

A friend in the College of Forestry, a scientist, says I've got it all wrong. The logistics of the project are faulty to begin with, he says, and that makes him mad. The logs will snap off the cables under the pressures of spring runoff, for example, and there are easier, less intrusive ways of achieving the same ends anyway. He's gritting his teeth as he explains this to me, trying to be patient. What really bothers him are the huge scientific egos involved in the project, and the millions in government grants. Can't I see the arrogance implied?

I'm easily impressed by science, I'll admit, but I'd still argue that there's something admirable here, however mixed the motives or flawed the technology. The designing of the stream is admirable because it's trying to put things right. It's trying to put things back, the way they were before the dams and the fishermen and the four-wheel drives. There's a kind of necessary humility in that act, it seems to me, a necessary reverence, or at least that has to be one eventual effect. The designing of the stream is not an effort to impose a convenient structure on the natural world but an effort to understand the complexity and chaos and randomness of what was already there before us, and of course that effort must ultimately fail. We can never get it entirely right. Nature will exceed us, the complexity of the stream elude us, and even if egos were big in the beginning the scientists must come away abashed.

And maybe calmed, as I was. Heartened. Trying to design a stream has to involve the contemplation of water, the observation of the endless flow—and of the bending of the alder and willow, and of the leaves moving in the wind, of the sky overhead. It has to involve looking, trying to understand, and that has to lead to hours sitting on the bank, clipboard or not, watching how the water moves, staring at the rocks, seeing leaves fall.

I'm all for it. I think the government should try designing as many streams as possible, all over the country, improving the technology as best it can. It couldn't make things worse—could even divert us from more damaging concerns. It would give jobs to the unemployed millworkers. And even if the projects failed for fish or the arrogant came away unchanged there would still be some P-5 pools and the alder leaves shining like pennies. There would still be the riffles and glides, and the sound of water, the endless rushing and flowing and eddying of water, beneath trees.

Deeper in the Forest

We would fain take that walk, never yet taken

by us through this actual world, which is perfectly

symbolical of the path which we love to travel in

the interior and ideal world.

—Thoreau

When I first completed my study of the research harvest and the profile of McDonald Forest, I thought I'd go on to write a local history. It would begin with the history of my own house and piece of land, radiate outward to the Willamette Valley, Oregon, and the West, and include all the stories I could find about the native Kalapuya, the settlers who came on the Oregon Trail, and the founders of the city. I thought to dignify my life by seeing it as history—history in the sense of story, of drama. Maybe a mountain man once walked the ground where my children now leave out their bikes and their squirt guns. Maybe this hillside of lawns and roses was the scene for some ancient Kalapuya rite. Maybe our apparently shapeless, inconclusive lives are in fact the latest moment in some large historical progress, caused and explicable.

But the longer I studied the old survey maps and pioneer journals, the more the picture blurred. Hours would pass in the rare book room or at home among the Xeroxes and still there would be no clarity, only rain and sun and the endless motion of the oxen, only the hills of grass and the forests going by. There was no defining action, only scene—only the minutiae of lives, the mud and the dwindling provisions, the sky and the sound of wind. "It is one of the attractions of the unknown," Virginia Woolf says in "Lives of the Obscure," "their multitude, their vastness; for instead of keeping their identity separate, as remarkable people do, they seem to merge into one another." The lives I found were obscure—Sarah Cummins, Bushrod Wilson, Lester Hulin—and I couldn't keep them straight, couldn't resolve their detail into pattern. But gradually that began to delight me, to lull me into some other, deeper rhythm. I was diffused into atmospheres. "Nothing much happens," Woolf says, reading old memoirs and autobiographies in some dusty Victorian library. "But the dim light is exquisitely refreshing to the eyes." Now one detail stands out, now one scene, but then, the silence of books all around me, the smell of yellowing pages and copying ink in my nose, "gently, beautifully, like the clouds of a balmy evening, obscurity once more traverses the sky, an obscurity which is not empty but thick with the star dust of innumerable lives."

I found plenty of evidence, too, for the first kind of history I wanted, proof that mountain men did walk where the bikes now lay, leading their mules over what is now asphalt and bark dust. This is when I learned that the mud and marshes of the Willamette Valley forced the first trappers of the 1820's to establish the California Trail in the hills just above the valley floor, where my house now stands—people like Alexander McLeod of the Hudson's Bay Company and the great Jedediah Smith. This is when I learned that the Applegate Trail of the 1840's and 50's, the southern spur of the Oregon Trail, went right past us, too, a little farther down, the Conestoga wagons and the dusty oxen rumbling along near what is now Highway 99. Levi Scott, one of the founders of the trail, homesteaded no more than a mile from here. Jesse Applegate himself lived not far away, in Rickreal. All the first Donation Land Claims in this part of the state were filed in these hills, above the muddy valley—Thomas Reed's, Abram Fuller's, David Carson's. And before the trappers and settlers, for maybe ten thousand years, the Kalapuya walked the ridgelines and camped along the streams, burning the forests season after season until there was only oak savannah, telling their stories over the long, rainy winters.

From my own little acre and a half I could have witnessed a part of the great Western Migration, seen for myself, firsthand, a piece of the history glorified in all the movies and books. At least some of the people who came by here must have done a heroic thing, or were about to. It's not impossible that some great sacrifice or struggle happened even within earshot.

But the impression you get reading all the old journals and letters is not of heroism or adventure finally but of hard work, of day-to-day complexity. "It was not a spectacular but a laborious history," Bernard DeVoto says in *The Year of Decision: 1846*. The wagon trains didn't march in a straight, unerring line but "strung out along the trail aimlessly, at senseless intervals and over as wide a space as the country permitted." Consciousness "dwindled to a point" as the wagons rolled on. There was "great strain and aimless suffering," great "perplexity and indecision." Pioneers were often "sunk in lethargic despair," were often lost for days at a time.

In 1914, in her *Autobiography and Reminiscences*, 91-year-old Sarah Cummins recalls the drudgery of the journey west:

> We continued our daily journeying, listening to the regular tramping of the poor four-footed beasts over the plain and through the dust, the midsummer sun beaming through the cloth roofs and the look of stern desperation setting in the countenances of the most refined and self-sacrificing.

She recalls the confusion of the wagon train at the sight of Sioux warriors:

> The entire company seemed almost wild with excitement, children crying, mothers screaming or praying, men running wildly about, not knowing what to do.

In a long chapter, "We Cross the Cascade Mountains and Are Lost for Eleven Days," she recalls the "slow and toilsome" effort to bypass the dangerous Columbia River crossing at the Dalles and reach the Willamette Valley overland, through the mountains over Mount Hood:

> Heavy fall rains were coming on and the steep slopes were almost impassible for man and beast. On the sixth day we

became entangled in a thicket of vine-maples and were compelled to turn back to our camping place of the previous
night. Next day we found it impossible to proceed through
the dense growth of the Mountain Laurel. . . . The old gentle-
man, Mr. Carson, had been chosen guide and he was misled
by the Indian trails that led to the berry patches far up on the
slopes of Mount Hood. So we had been making little progress toward the place of our destination. One morning we
awoke in a blinding snowstorm. We toiled along the whole
day through without seeing a tree or a spear of grass. Our
course seemed to be up a gradual steep slope. As night was
coming on it seemed we must all perish, but weak, faint and
starving we went on.

Wet with snow, stumbling from exhaustion, the party finally
stops to camp for the night and struggles to make a fire:

Most of the men and all the boys were shedding tears. Not a
man could be found whose hands had strength to fire a gun,
and not a thread of clothing for kindling. All were panic
stricken and all hope seemed abandoned. . . . As soon as he
realized the situation my husband seized the gun and fired it
into the little bunch of kindling the men had prepared, but no
fire resulted. He now made every man present haul off his
coat and in the inner lining of Mr. J. Moore's coat a small
piece of dry quilted lining was found. This was placed in a
handful of whittlings, and as the gun was reloaded all realized that upon that charge depended our lives. With almost
super-human effort Mr. Walden succeeded in firing the gun
and in an instant the flames burst forth. A great shout of
thanksgiving burst forth and each poor suffering traveler
crowded as near as possible to the welcome fire.

This was the journey west, not the easy structure of story, not
character-conflict-resolution in one dramatic arc, but a fighting
through forest and brambles, a wandering over a mountainside,

a painful, pathetic effort to get a fire started. There was courage here but it was the courage of stamina, of dumb, plodding perseverance, and it all depended on the simplest things, on the sparking of the kindling, the drying of the coats.

Bushrod Wilson and his friend "Ritz" were lost for days trying to get to the valley from the mouth of the Umpqua River, on the Oregon Coast, where a ship had left them. "This was the forest primeval," Wilson says of the Coast Range, in an old, mimeographed copy of his "How I Came to Oregon, As Told at a Meeting of Oregon Pioneers at the Corvallis, Oregon Opera House, December 31st 1886." The expanse of forest and mountains appears "grand and appalling" and "as we thought of cutting our way for a hundred miles through it our courage all but failed us." The rain continues to fall, and again and again they try to hack their way through the underbrush. "When we got down to the river the brush on the banks was too thick to make any headway through it and the water in the stream too deep to wade." Each day they reckon they've walked fifteen miles or more but advanced no farther than three or four "on an air line." They keep "making out" their way up the hills, "often going one step and slipping back two, grasping the salal brush to pull ourselves upward, climbing over or going around fallen monarchs of the forest." Their pilot bread runs low, then their beans and rice. They find a spring. Ritz manages to bring down a doe.

The Oregon Trail was as much a forest as a path; as much a mountain range as a highway. And the people who traveled it stumbled and fell and occasionally got off enough lucky shots to survive, wandering finally into the valley, dazed and lucky, not sure where they were or if they even cared.

I don't mean that all heroes are fools and knaves but simply that the people of this history are people, as bumbling and vulnerable as the rest of us and so finally more compelling than heroes. I don't mean that there is no plot, no meaning to things, but that the shape of these details is subtler than history, subtler than story. There is an ordinariness, a tenderness in these documents that moves and instructs us not because it exceeds our lives but because it so deeply resembles them.

Lester Hulin came to the Willamette Valley on the Oregon and Applegate trails in 1847, keeping a diary along the way in a 6-by-8-inch leather-hinged day book. The entries scrawled on the faded marbleized paper are brief and halting, and yet almost lyrical somehow, their roughness a kind of poetry.

Oct. 1st   Passed around a lake about 10 ms and camped on a small cool stream

Sun 10th   We found pleanty of water for 5 Ms today we should have come here to camp, but did not so we did not make the next camping but took a ridge in the timber and found a small opening with good grass but no water   camped here   distance 16 ms

T 13th: Continued on over the Mts; through the timber in about 8 ms we descended a steep hill to a creek up to the top of the Rogue river Mt then down for about 2 ms   The valley increased in width and the face of nature became more interesting   during the day several Mt. branches had increased the main stream considerable; at noon we saw some indians and their lodges or shanties   they ran like wild men from us; as wild as the deer they seem

One evening the wagon train is "visited by savages" who "vent their spite" at finding no cattle to steal by shooting three arrows at a young woman baking bread over the fire. One hits her in the calf, one in the arm. The next day the men hide in the willows waiting for the Indians to pass but only hear them "halloo" in the distance.

And yet it's not adventure the day book evokes, it's weather and distance and topography. The wounding of Ann Davis is mentioned only briefly, in passing, entry after entry recording the number of miles traveled, the quality of grass and water. Lester Hulin was not an educated man, and language can never convey the fullness of experience. And yet reading his day book you get the sense that this is the way it really was, the scene dominating all the actors in it, the landscape filling every consciousness. You get the sense of progress so slow and distances so great the present seems eternal, the ox frozen against the yoke, the forest and fields everywhere.

The day book ends November 5th, 1847, with the entry "moved down about 4ms and camped   day rainy." There's a space and then this inscription, written, apparently, years later:

Thus passes the journey of life   day after day
glides swiftly on and life is but a Span

And you think, this, too, is the way things really are. This is how time seems to all of us. The journey west is finally over, the Donation Land Claim filed, the 320 acres worked and worked, corn coming up, and beans, the cattle grazing in the upper pastures. You marry, the children come, and life passes swiftly by, rich and full and occasionally joyful yet not a story somehow, not an adventure, not a plot you can get into words even if you wanted to. The last sixty years of your life are a gap on a page. The journey on the Oregon Trail was slow and torturous and the life in the valley swift and happy and yet both have the same structure: each day always in the present, the succession of days like a fullness, a richness seeping out, spreading.

"A person's life is not a series of dramatic events for which he or she is applauded or exiled," Gretel Ehrlich says, "but a slow accumulation of days, seasons, years, fleshed out by the generational weight of one's family and anchored by a land-bound sense of place." That's what Lester Hulin knew, I think, and what his day book manages to express: it is a book of days, of the slow accumulation of days.

The Kalapuya seem lacking in history, too, in the shape and cogency of story, though they were so ravaged by the malaria and smallpox the trappers brought in the late eighteenth and early nineteenth centuries that their culture had all but disappeared by the time of the first settlers in the 1840's and 50's. Unlike the Plains Indians to the east or the Coastal Indians to the west they never seem to have been warlike. Their culture was without hierarchy and specialization, they produced no brilliant beadwork or complex weaving, they threw together simple shelters of fir and cedar branches whenever it rained or merely waited out storms beneath the trees. There are beautiful little agate arrowheads scattered throughout local museums, and larger projectiles of quartz and obsidian, the surfaces delicately scored and worked, but few other artifacts have survived. The linguist and ethnographer Melville Jacobs is almost apologetic about the spareness and understatement of the myths he recorded. It's as if the Kalapuya blended into the mists and rains of the valley, merged into the softness of the land. Without the challenge of harsh climate or difficult terrain they simply hunted and foraged and moved through the tall grasses, generation after generation, gathering in the evenings in the willows along the river banks or beneath the oaks on the open hillsides.

The myths and stories that Jacobs preserves in *Kalapuya Texts* describe the way the world was made and account for the cycles of nature. They seem oddly rambling and unresolved, often scatological. "The News Precedes Coyote" tells of Coyote going along the Willamette toward what is now Oregon City. He makes camp, it gets dark, he sleeps and resumes the journey the next day. One evening he transforms a sweathouse into a large, cavelike rock so that he can lick himself in private during the night. The next day, seeing some people in a canoe, he calls out, "What is the story? What story is there?" and the people reply, "There is no story at all. The sole story is that Coyote was licking himself last night." Taken aback, Coyote returns to the rock and sees the large crack where the people looked in at him. "Oh I suppose that this is where the story came out from it," Coyote says of the rock. "That is how it is going to be, that is the way it will always be. No secret will ever be hidden."

The secret of the Kalapuya seems to be that there was no story at all, just the succession of days, just the weather and the sky and the telling and retelling of the myths. And more and more I found this compelling, this quiet, this gentleness, however different from the kind of drama I was hoping for at first. More and more I began to feel a quiet, a peacefulness beneath the terseness of their speech, especially in what Jacobs calls the "Ethnological Notes" preceding the myths themselves:

Long ago when the people saw the new moon they spoke to the moon. They said to it, "We are still alive here yet. We see you now that you have come out again, and we are still alive here yet."

Long ago the people observed everything with care.

Long ago the people used to say, "When the large-owl hoots near your house, when it sort of sobs, he is telling you that you are in some sort of trouble."

Long ago the people would say, "Now the wind is blowing hard. Now those firs are dancing."

Life moves in its rhythms and its cycles, the seasons change, the rains come and go, and everything is observed with care. The firs dance in the wind and your heart leaps up. The new moon rises in the sky and the beauty of it moves you to speak. You address the moon, declare your own life to it, and that moment of insight, of speaking, is separated out from all the other moments, lifted out of the cycle of things.

"Always a boy who wanted to become a shaman, he was always swimming in the early morning," another "Ethnological Note" begins. "And when it became dark and the moon was full, then he would go to the mountain. He would fix up that spirit-power-place on the mountain. He would go five nights. And always in the early morning he would be swimming. And then he would find his spirit-power, while he slept he would see his dream-power, his spirit power." In a dream you are given a glimpse of your own true self. Always in the morning you are swimming, and as the water slips past your arms, your legs, the very presence of the world, the simple fact of things seems to flow all around you. You climb a mountain and fast and pray and you come away changed, though you don't know how exactly, can't explain who you now are. Do not be fearful on the journey to the mountain, the people say. "You must get to where you have gone to."

Sarah Cummins and Bushrod Wilson and Lester Hulin climbed their mountains, too, and from the summits they looked out at the beauty of the valley.

> The sky was cloudless. The storms of the previous day had so cleared the air of dust and impurities that my horizon was boundless, and this, my first prospect of everlasting green forests and their wonderful vividness, green on all the near approaches and changing with wonderful blend from green to ethereal blue, and on the distant margin rested the shade of blue, so intense, so indescribably beautiful that no power of words can express the wonderful panorama of beauty with which my soul was entranced.

After all the days and weeks of mindless effort and drudgery, the hacking through the forest and the losing of the way, Sarah Cummins comes to the summit and the expanse of sky and the panorama of the valley and for a moment she has a vision of the world whole and boundless. Bushrod Wilson and his friend Ritz, grabbing fistfuls of salal as they scramble up hill after hill, falling and stumbling and lost for days, suddenly break out into the open:

> Just before we reached the summit we came out on a beautiful open plateau covered with luxuriant grass in which deer were feeding. They did not run away at seeing us but gazed at

us with timid indifference. . . . It was probably a quarter of a mile as across thin clear space and the grass on it was soft and velvety. We crossed it and began the final ascent to the summit. We were now above the tops of all the surrounding mountains. . . . Between us and the mountains and not very far away, there stretched a beautiful valley and we could see stretches of several canyons that seemed to run in an easterly direction and we did not doubt that we would soon come to some creek that would eventually discharge its waters into the large river, parts of the surface of which could be seen. We looked back over the country we had just crossed, a jumble of mountain tops covered with evergreen timber but away beyond the mountains where their tops seemed to touch the horizon was a band of trees and then the blue horizon.

After all the jumble of the mountains and the bewildering tangle of fir, finally there is this, the wide sweep of the valley and the sudden flowing of waters, and even poor, prosaic Bushrod Wilson, sailor and carpenter and soon to be homesteader, feels himself lifted up to a kind of lyricism. He looks back on where he's been, at the mountains, at the blue horizon, and for a moment it all makes sense, seems to have a kind of deep coherence. And before him is the valley, the green and beautiful valley, where he soon will descend and live out the rest of his life.

There are these moments, too, then, not just the gliding past of the years, the passage of time "without conscious measurement" as Sarah Cummins puts it at one point, but also the brief glimpses of something larger and more moving still. There are these moments, too, which do not advance the plot, in which we move no farther west, in which, in fact, we halt, step outside the story, outside of time altogether. For a brief instant everything is suspended and we simply see the world spread out before us, simply look out at it with gladness and recognition.

All history is local history. All history is *located*, in the one given moment, and then the next, and the next.

One of my early memories is of the day President Kennedy came on television to warn the country about the missiles in Cuba. We lived on the Milk River, outside of Malta, Montana. It was October and the cottonwoods were turning brown, and the broad, sluggish river behind us was brown, too, swirling and eddying. Some older boys had dragged a davenport to the bank and were fishing there, sitting down. I was forbidden to go to the river, but I'd been down there anyway, watching the boys, and had just come up to the house.

The screen door opening and slamming. The coolness of the livingroom after the warm sun. My parents listening to the president, his image flickering.

The week before Kennedy was assassinated I'd shanked a punt in the Kick-Punt-and-Pass contest and earned only 94 points, after weeks of practice in the backyard. I was still moping the day of the funeral, kicking up leaves in the gutters, but sometimes I'd come in to watch TV. I remember seeing little John salute.

I remember slipping and sliding on a frozen pond that winter, us boys in parkas with the hoods up, holding onto Mom and trying to keep her from falling, the prairie all around us, the sky. We're laughing. Christmas I sang the part of the Third King. July a thunderstorm blew down the drive-in. Halloween a cocker spaniel nipped my palm through a chain link and Dad had to come get me in the Jeep.

I went to high school during Vietnam and I don't believe I ever gave the war a moment's thought. It was given, part of the scenery, and in any event had nothing to do with me. We were living in Spokane by then, in the house on Lincoln. The war appeared on the screen in the livingroom every evening, images of violence and horror, flat, steady, and images of protest, anger, social convulsion, and looking back I realize that I was only two or three years too young for all that, only two or three years away from my own eligibilities. But the war passed me by. It never penetrated. My adolescence was a time of pleasure and instinct and motion and bursts of color and light and everything coming easily, naturally.

What I remember are summer bike rides down the corridors of trees to Alan's house or Shadle Pool, riding in a reverie, slow and meandering, coasting through patterns of leaves and sun and cool shadow. The afternoons were deep and silent. Sprinklers oscillated on lawns.

Later I took long winter walks on Five Mile Bluff, brooding on the rocks, looking down on the city and indulging my pleasant miseries. The wind was blowing in the ponderosa, and I imagined myself cutting a romantic figure, outlined against the sky.

I don't remember the Tet Offensive or Kent State. What I remember is taking Alan's paper route for a week and the stars at 3 A.M., the wet lawns, the smell of lilac on front porches. What I remember is hammering on a science project in Randy Cade's basement and walking home in the snow. Bright cold, white clouds in blue sky. What I remember is Jerry Evans and me sneaking under the fence at Cunningham Sand and Gravel and screaming off the sandy cliffs, flinging back our arms, pedaling for distance. We're suspended in the air. And then the umph into the soft, forgiving sand, and the climb back up, and the second run, and the third, and the sky growing pink and orange. The leaping out and landing. The leaping out.

One summer the Scouts spread hay beneath the ski lift on Mount Spokane, and pausing in my work on the side of the mountain, the sweet smell of hay in my nose, stalks in my hair, my pants, I looked down on the softness of the world, the green and distant valley, and my heart leapt up at the beauty of it, like a fairy tale, like a story.

Getting older I've had to learn to think more abstractly and politically, as much for selfish reasons as out of social conscience. I know that distant things can hurt me. Aware of all the systems and ecologies I belong to, all the balances, I sit in front of the evening news, like my parents in the sixties, and feel a growing dread. I hear the latest reports on the greenhouse effect and think of waters lapping at the foot of the Coast Range, here in Oregon, where I live now, raising my own family—the fir trees migrating north and the dry land blanching in the hills behind us. I hear of more tension in the Gulf or the fall of the yen and I think of the future these facts might foreshadow, gas lines, maybe, and empty suburbs, my children out of work and struggling.

But even now I can't help generalizing from the present. Against the backdrop of my larger worries I find myself still heartened by the beauty of things. Saturday mornings I sit in front of the picture window in the livingroom, and I read that the Iraqis have lobbed more Scuds or that Gorbachev has been deposed or that the Serbians are advancing on the Croatians, and I stop to think about all that, to worry. And I do. Yet looking up from the paper I am always distracted by the weather and the trees, by the wind in the fir, say, or the clouds breaking and the wet-washed look of the sky, the yard filling up with light. I can't help taking pleasure in the birds at the feeder and on the branches of the magnolia, bickering and jockeying for position, blurring at their perches, pecking and jerking for seed, then squirting off across the yard, like swimmers, diving and streaking in the morning air—cartoon birds, little fantastics, clowns— mountain chickadees, dusky-backed chickadees, black-caps, red-breasted nuthatches, Oregon juncos, scrub jays, pine siskins, towhees. There is music playing. The kids are cleaning up their rooms. I am drinking coffee. Later we have lunch and return our books at the library, emptying out the baskets and stocking up for the next week.

The wars and the chickadees are both on my mind at the end of the day, in conflict, the one shadowing the other, but it's the chickadees, and the light, and the trees that always feel most real and important, that I always seem to remember.

I think again of all the meetings I attended when I learned about the patchcuts and the clearcut, a slight weakening in my knees, my mouth dry. The environmental debate was right on top of me now, unavoidable. That's what I felt then. Trees were going to fall, and I would hear them. Roads had already been "improved," trunks along the trails numbered and beribboned. "Site productivity enhancement," someone said; "multiple-output commercial asset," someone else. And I sat there on the folding chair, sick at heart.

But that's been several years ago now, and as it turns out I can't really see the cutting at all, unless I climb the road. The clearcut only shows from the top of the hill, and the smaller patchcuts have been folded into the larger forest, invisible in the far view. All I see from my window is my own familiar wall of fir, the illusion of a forest extending off into the distance, and more and more that seems to content me, despite the letters I've written and the protests I've lodged. The birds still stage their runs there, waiting for a chance at the feeder. The light still hits the top branches in late afternoon. The slow peace of evening still descends.

The kids and I often walk to the clearcut, and though their favorite trail has been destroyed—just open ground now, stumps, a tangle of brush for slashburning—they don't seem to notice. They play with the sticks and walk on the logs, swinging from half-fallen branches. I'm glad that like me at their age they haven't yet learned to worry or grieve. They accept the fallen trees as given. Everything is a stage for their play and exuberance. Maybe what they'll remember in the years to come is walking in the woods with their father, counting the rings on the big logs.

I don't entirely fault myself for my own satisfaction in the chickadees and the light. We all need ways of handling the things we cannot change. There is some little piece of the human spirit in this enjoyment of the particular, this celebration of what we touch and taste and feel, of our own given lives. What I worry about is complacency and parochialism. I keep thinking of how many problems have worsened and injustices continued because people couldn't imagine what was going on beyond their "immediate viewshed."

How close would the cutting have had to come to make my grief more permanent? How deep must a forest be to satisfy the eye?

One night last week I woke up suddenly, adrenaline racing. Moonlight flooded the room, silvered the trees in the yard. Sirens were approaching from a distance, the sound of engines. Closer and closer, and in the silence I could make out the details of a chase: two, maybe three police cars in pursuit, slamming around intersections, and a smaller car, tires squealing, rpm's racheted up, whining, trying to get away, zigzagging up the country roads toward us: Lewisburg—Mountain View—slowing—accelerating—the little car—the sirens. Overlook! And I could hear tires leave the road at the top of the rise, and the heaving of a chassis as it hit the ground—could almost see the sparks—and the red lights of the police cars are reflecting on my bedroom wall, and for a moment it seems that all those lines of sound are intersecting at my house, that the chase is headed here, for my driveway, and I think I should jump up and check the doors, run downstairs to the children's rooms.

But then the cars streak past and the sound recedes over the next hill, and the next, and I am sitting bolt upright in bed looking at the light of the moon. My heart begins to slow. A digit on the clock flips: 3:05.

The moon is so bright the lawn seems like snow. The moss on

the oak limbs is phosphorescent. It's as if the scene is photographed through some negative lens, and we are all beings of
light, milky and pure, blurring as we move. I survey the room.
The philodendron on the dresser and the statue of Mary. The
orange crates full of books. The louvered closet doors. The roll-
top, my cords crumpled on the chair. Barb's slippers. Through
the vents I hear my son snoring.

I fall asleep in the light of the moon, and the trees glow and
move in the wind, and the owls come and roost in their branches.
Our cat prowls in the tall grasses, lifting each silvered paw.

Several weeks ago a friend and I drove up to Seattle for
the annual meeting of the National Council of Teachers of English: five thousand name-tagged English teachers milling around
book displays or sitting in close, thick-carpeted conference
rooms listening to papers with titles like "Ideology and Praxis
in the Postmodern Classroom" or "Sentence Combining Revis-
ited." We spent the first afternoon walking along Lake Washington, talking about our lives and looking at the houses crowded
on the shore, glimpsing interiors. The Burke-Gilman Trail lay in
shadow, banked by tangled blackberries. A seaplane skidded on
the brilliant lake. Mount Rainier hung over the Naval Air Station. At Matthews Beach I showed my friend the tiny house
where Barb and I lived during graduate school and all those
memories welled up of living our own lives in the wake and
shadow of the city. The next morning we wandered through Pike
Place market. Bok choy and brussels sprouts gleamed from racks,
beaded belts and fine-tooled leather and bright embroidered
cloth dangled from booth after booth, the famous fish throwers
shouted and performed near the main entrance, twenty-pound

coho flying from the display to the scales, people everywhere, the smell of espresso, Swedish pastry, wet wool. The weather was gray. At the window counter of a little cafe we ate eggs and hash-browns and watched the ferries come in.

It's not that the concurrent sessions back at the Conference Center were unimportant. I admire scholars, more and more since I admitted I'm not one. It's a question of natural tendency, and in a sense, of specialization. Somewhere a few years ago I stopped checking my instincts and let my mind work the way it works and more and more I think I've entirely lost the knack of identifying issues, making arguments, and joining the debate. I've veered off toward the particular, the peripheral, accelerating so fast in that direction that sometimes I feel myself lost in a kind of aphasia, my left hemisphere almost dormant. All I'm inter-ested in, all I seem to process, are moods and moments, atmo-spheres, sudden scenes and faces. I'm swept along at the market, dazed, sometimes elated, and the fish are flying around me and the ferries are coming and going and sometimes the mountain appears from behind the clouds and sometimes it doesn't.

I've forgotten most of the theories and terms I learned in graduate school.

I don't read the professional journals and reviews.

I prefer most literature to most literary theory—just about any poem, play, story, novel, autobiography, biography, and history to just about any critique, analysis, and treatise.

And yet I never remember plots and characters. I'm devoted to Raymond Chandler, for example, but without a feel for the famous intricacies of his plots, any memory of his murderers. I read for atmosphere and light, for flashes of scene and charac-ter—the DeSoto pulling out into the rainy street, the smell of eucalyptus, Philip Marlowe in shirt sleeves making coffee in the kitchen.

Lately I've plunged into the Civil War, especially Lincoln bi-ographies, listing battles on the inside covers, in order, with their generals and who won. But the names and dates soon blur. What moves me in the histories is the bare trees and steel gray skies, the smell of gunpowder and campfires, the sense of weariness and cataclysm. I'm thrilled by any glimpse of actual lives—the rattling of a mess kit, laughter. I'd like to know what Lincoln ate for breakfast and how he folded his long body into a chair and I'd like to hear the high-pitched voice telling backwoods stories or asking for a glass of water.

Today I've been reading John Cheever's journals, and what amazes me here, as in his stories, is the grace and capacity of his descriptions, that invoking of a real, tangible world, so vivid that it overpowers (for me) the awareness of his alcoholism and bisexuality and the moral dreariness of his characters' lives: "How subtly the air becomes fresh at dark and how a perfectly round, pale moon comes out of the woods. There is the excitement of autumn in the cool damp air and the light of the moon, coming back across the field through the orchard, the rich smell of windfalls, the beautiful flavor of an apple." Why this listing of details works on me with the force of revelation I can't say, exactly. Cheever thought light itself was somehow "salvific," and that's part of what moves me in his writing, I guess, the light pouring out of rooms or fading behind mountains, the movement of a hand under a lamp, just the sun coming up in the suburbs, firing the rose bushes.

My reading has become purely "literary" in the sense of dwelling in the kind of visceral world only literary detail can re-create, the impulse for meaning and extraction suspended. I sit in the easy chair in the livingroom, in front of the maple tree and the picture window, the trees all around me, and I drink coffee and read and the afternoon fades to evening and I reach up to turn on the light.

Later I can't find the dog-eared page, or I wrote the quote down but left out words or page numbers, or I can't find the slip of paper. I've shelved the book out of order or left it under a pile of magazines and when I stumble on it in a year or two and glance at a page, start rereading, the story comes back to me like a forgotten place, a surprise, and I am caught up in it all over again, the place, the world, reading without a pen and paper, the light of another day fading in the room.

I keep thinking of the harvest, how it shocked me out of passive watching, mere seeing. The sudden falling of the trees demanded some response, and that meant learning the jargon of forestry and the new paradigms for understanding the ecology of things. There were letters to write and an administrative structure to influence. The challenge I felt was the challenge of stewardship, of political responsibility—a pull outward, into the world of fact and negotiation.

But then reading the paper one morning—the first year or so had passed now—I realized I'd been skipping over the latest reports on the spotted owl and disputed timber sales, guiltily bored. The rhetoric ran together. I started finding excuses to miss the weekly meetings of the Sustainable Forestry New Paradigm Working Group, pleading busyness. My ecology texts lay facedown in the corners. In quiet moments walking in the forest or sitting in my easy chair I found myself drifting back, being pulled back, to some familiar center, some inevitable inwardness.

Last Fourth of July Barb and I led the kids on an expedition to the top of the little mountain behind the house to see the city's fireworks. John carried the apples, Maggie the popcorn, and Timmy held Max's leash, in case we encountered other dogs. It was dusk when we left but twilight by the time we got to the top and the lookout was lonelier and scarier than we imagined. We had a fine view of the valley's lights, like on an airplane's final approach, from Monmouth in the north almost all the way south to Eugene, but the wind whistled and the stars came out and we felt a million miles away, chattering to fill the silence. We sang songs, shone the flashlight at constellations.

Always at night in a forest or beneath a starry sky there is a sense of expectation, of something about to happen, so when the first volleys burst and sprayed over the river the kids had already had enough and were begging us to take them home. It's marching down the hill and back into the forest that stays with me now, that image of leading the kids through the trees (behind us, out of sight, the explosions and colors of the fireworks), the tall trunks swaying and groaning. I could hear Barb smiling in the back of the line, reassuring Timmy, and I could hear the fear in the children's voices. The forest seemed like a presence all around us, palpable. We were deeper in it than we ever had been. We stopped to shine the flashlight up the trunk of a fir, and for a moment it was illuminated, sweeping upward, gray and ghostly, twice as tall in the darkness, and above it the stars, the tossing of the other trees.

Nothing happened in this moment in the woods, nothing tangible was gained, and yet even as we moved down the trail back to the house I was trying to hold onto it in my mind, grasp it and hold onto it and follow it to its source—to the exclusion, for days, of anything else. I had to stop and consider it.

Why am I always drawn back to scenes like this? What knowledge is disclosed, what meaning revealed?

What are my obligations?

Just yesterday, talking with the Dean of the College of Forestry, I was bewildered all over again by the complexity of the environmental debate. The population of the world will double by 2030, the Dean said, from 5 billion to 10 billion, and those people will need some kind of raw material to survive—if not trees from the Northwest, trees from somewhere else, perhaps Third World countries where laws are much weaker; if not wood fibre, some other source of material, concrete or plastic or steel, material more environmentally damaging to extract and process (the Dean said) than any method of growing and harvesting trees. There are layers and layers, complexities and economies, and the forest I live by participates in them, is a site, like every piece of ground, every part of the earth, for management and decision.

I should have challenged the Dean's assumptions, I realized later: why not first try to eliminate waste, change the way we live, prevent the population explosion to begin with? But I didn't think of these questions at the time, and in a way that's the point.

In the face of these problems—this powerful, convincing language of fact—what value is there in a walk in the woods and the moving of trees in the wind, the night stars, the look on my children's faces?

Why can't I keep my mind on the world outside of me?

Whatever piece of paper I put in a file I immediately lose track of. Nothing in a file folder ever gets looked at again. If I actually want to find and use them I need my papers and notes and books scattered in piles on the desk and the futon and the chair and the floor, at least a corner visible, and even then I usually don't stop to search but just paraphrase from memory.

In the office I throw letters, memos, brochures, or student papers I don't have to act on immediately into a single metal basket and at the end of each term throw the whole pile into the recycling. Each day I stand at my mailbox and directly toss anything without a stamp.

I forget to write down appointments or forget to go when I do or schedule two for the same time or misplace the appointment book.

I have a curious kind of localized amnesia when it comes to teaching. Right before I walk into a classroom I can't for the life of me remember what I'm supposed to do, how to get people to talk, what I'm required to know, how exactly the necessary give and take is managed, as if I've never done it before. And right afterward I'm blank, too, unable to recall in detail what happened. Yet in the room itself, during those fifty minutes, my mind seems to open up, students respond more often than not, idea follows idea and I'm sometimes exhilarated by their clarity.

I seem unable to function anymore in committee meetings. I have no idea what people are going to say or why and I can't seem to formulate any statement that will move the group in any certain direction. I'm mystified. What are we doing? Why are we doing it?

I have an increasing reluctance to talk to more than a few people.

I have an increasing tendency to make ad hominem reductions, rejecting the ideas of people I don't like, accepting the ideas of people I do, regardless of the quality of the ideas.

I'm extremely susceptible to handshakes, touches on the shoulder, small talk, the speaking of my name, casual invitations, even superficial courtesy.

My handwriting has become inexorable, lowercase letters blurring into lines, s's sideways, e's opened up like u's.

I've always had a kind of manual dyslexia, an absolute inability to assemble parts or puzzles or tie knots or fix anything and now it's getting worse. I'm just stone stupid about machinery, missing the gene, and I don't try to cover up anymore.

I hate rearranging furniture. Moving an easy chair a foot in can keep me up at night.

I hate writing checks, stopping to take out the book, trying to find a pen, looking over my shoulder at the people waiting.

I hate having to stop and gas up the car.

I hate driving into town on errands.

I hate being rushed. In the morning I need to sit and stare for at least half an hour, and then I can't focus on the kids and the hecticness of everyone trying to get out of the house simultaneously. Barb has to supervise, juggling the phone and the calendars and the lunches and the socks and the car pool arrangements in the afternoon, football practice, piano lessons.

What all these quirks and tendencies have in common is a serious failure of patience, an increasing aversion to systematic thinking—any slow, sequenced consideration of processes and objects in the external world—and a growing unwillingness to be distracted from some deeper, inner work. A radical specialization and narrowing seems to be going on inside of me, as if I have to gather all my energies for some single internal project and so everything else on the outside is suffering.

Sometimes this narrowing fills me with self-loathing, sometimes a kind of detached fascination, and in any event I feel it getting more and more intense, developing its own momentum. My other voices are fading, things are falling away, and I'm not at all sure who I'm on the way to becoming.

A weekend at the beach without the kids. Lent. Windy and mild. Barb and I stay in our room watching old black-and-white movies and now and then take walks on the beach, picking our way over driftwood.

Sunday we stop for mass at Sacred Heart in Newport, a little wooden church between used car lots on Highway 101. Inside smells like pancakes and eggs. We can hear the cars whizzing by on the highway, the campers and the RV's. There is a life-sized wooden crucifix over the altar, Christ hanging down his heavy head, his chin on his chest, and from my angle in the back of the church the crown of thorns seems for all the world like swimming goggles. A crown of goggles. Christ hanging his head amid the RV's and the used car lots, and the priest with his hair slicked back, saying mass over the traffic.

On the way home we stop at Seal Rock for breakfast. The light is blinding now, the sun is out, falling through the restaurant windows, on the tablecloths and the spoons, our eyes filling with tears.

And there are seals in the waves, diving and bobbing their way alongside us as we take a final walk up the beach. Foam washes the blue-gray water and the trunks of the waves glisten like cut glass. The seals bob up quietly, two, three of them, moving up and down with the tide. We don't notice them at first, but then they're following us, turning to look and then flipping over, coming up again farther down. They seem nonchalant, unconcerned. Their heads are marvelously sleek. Sun hits the tops of the waves and mist breaks up over the forested cliffs and in the distance is a sail. We're walking hand in hand and the seals are floating in the waves, moving up and down, and we seem to be floating, too, just as delighted, moving with the tide.

Just this, whatever "this" is.

This is the kind of moment I don't want to be distracted from, the kind of thinking and remembering and considering that sometimes seems supremely self-indulgent and other times more important than anything else.

Once in the old house, when Timmy was still a baby, moonlight shone through the livingroom window. The rest of the house was asleep, the furnace kicking on at intervals, and I was cradling Timmy in my arms, giving him his bottle. Moonlight flooded the room. We were bathed in it, and when I reached up to wipe the formula from his mouth I realized it was moonlight spilling over his face and hands, moonlight in his hair. We were bathed in the milk white light of the moon, the night all around us, and the light was pouring on me, too, splashing all over my bathrobe, my bare legs, overflowing.

Why is this moment so important to me? Why when it happened did it have the force of revelation, of truth? What truth was revealed?

Sometimes I feel pulled and pulled down a long corridor to a single room. It's a pleasant room with a comfortable chair and a wide picture window and through the window is a beautiful forest, hill after hill of fir extending off into the distance.

Sometimes I think that the end result of the momentum I feel will simply be a silence, a moment when I'm finally not saying or doing anything, and sometimes that exhilarates me, and sometimes I realize that there's nothing I can do about it anyway.

IV

I'm paging through the *New Yorker* one Sunday afternoon and come across this paragraph. It's the introduction to a profile of a classical guitarist.

> By the age of thirty-seven, the ground of a person's life is more or less sown. The task of tasks has been taken up, ignored, or irremediably obscured. "Unwritten poetry twists the hearts of people in their thirties," the novelist Elizabeth Bowen once wrote. By the end of the fourth decade, we've detached ourselves from the mentors and stand on the bridge of our own story, looking forward and looking back. Already, some of us are faltering and losing heart. "I've done my best," we say, with a sigh. "The next, the rest, the others must carry on." With any luck, the artists are among the company that presses on; if not, one way or another, they begin to die.

I'm stopped dead in my tracks. *I'm* going to be thirty-seven in a month, and I keep wondering which category I've fallen into, whether I've taken up the task of tasks or faltered and lost heart. I keep wondering what right the author has to say such a thing and how old *he* is. Is any of this true? And at the same time I feel oddly honored to be old enough for the passage even to apply to me. I'm surprised. How did *that* happen?

A few days later we're sitting around talking after dinner, the Riesling half empty and french bread crumbs all over the tablecloth. I get up and get the magazine and read the passage to my father-in-law, who is sixty-five, and he just laughs, out loud, a long, pleasant laugh, and I laugh, too. The sun is showing through the magnolia, the prism is casting little rainbows on the ceiling, and my father-in-law and I are laughing over the wine and the dirty dishes, laughing at the idea of being thirty-seven and standing on the bridge of our own story.

I dream I've had a heart transplant. There's still a row of staples from my throat to my navel but I feel fine inside, snug and tight and fit, and I'm bouncing around from place to place telling everyone my story. I feel people should know. I'm still surprised about it all. I'm proud of myself.

I keep showing people my scars, describing the surgery and how light I feel, how hollow (I know this doesn't quite make sense). My chest cavity feels like a snare drum.

I dream that I'm wrestling my father in my best friend's front yard, near the curb. I'm the age I am now—it's my adult best friend's house, and his kids and mine are playing somewhere out of sight. Dad and I have been shouting, I've been telling him what I really think, not holding back anymore, and suddenly he says, "That's it!" and lunges for me, but I feint and go for his legs. I upend him easily. He's light, I realize. I could pick him up with one hand. I'm holding him down by the throat now and he's squirming, but I'm trying not to hurt him. I'm trying to stop the argument, and I'm hoping the kids won't come out and see us. "It's OK now, Dad," I keep saying. "Hush now. Every-thing's OK."

I've been reading for a couple of hours in the abbey li-brary. I'm in the back by the wall, in a carrel, hidden in the stacks. It's cool and quiet, the long rows of books surround me, and I've been reading about "Spiritual Practices and Helps" in the *Spiritual Exercises* of St. Ignatius, the sixteenth-century Span-iard who founded the Jesuits. "The *examen of consciousness* is the instrument by which we discover how God has been present to us and how we have responded to His presence through the day" (I'm reading). It's not to be confused with an "examination of conscience." The examen is, rather, "an exploration of how God is present within the events, circumstances, feelings of our daily lives." You just sit down at the end of the day apparently, after completing the more formal spiritual exercises themselves, and reflect on what happened to you, what you did and said and felt. Ignatius considered this final reflection even more important than the systematic scriptural meditations he made famous. If you couldn't do both the exercises and the examen, he thought, do the examen. It alone can sustain "one's vital link to God."

I've been reading for hours. I'm alone in the abbey library, and the words are forming and re-forming in my mind. My mind is like a page and my own thoughts have assumed the rhythm of sentences, quiet and steady, when suddenly, somewhere out of sight, I hear a voice. Someone has stopped to check out books. I can hear her saying her name to the librarian, and then I realize it's my wife's name I'm hearing, my wife's voice, and then I realize that's the one voice I most wanted to hear, the only voice.

It's Memorial Day, very hot. A breeze has picked up, moving the branches in the fir tree. I'm out on the deck barbequing fish when I hear the first spiraling song of the Swainson's thrush, the first song of the summer. It's almost evening and the light is low and the first Swainson's is back from Central America, singing in our forest, and I realize that I've been waiting for that sound, listening for it. All weekend I've been straining to hear the Swainson's thrush, staying ready, working outside in the garden or reading on the deck, and now I've heard it. I've not missed a single note. This is the first one.

Dan and I are going over the pictures he took on our bike trip to the coast. The five of us are standing in front of the coffee shop downtown, helmets in hand, bikes loaded and ready behind us. I'm coasting down the highway past a big billboard—or coming down off the hill by Logstown, trees all around, the highway slick with rain. Here we are posing in front of the ocean. It's been raining again, and the five of us are in our yellow rain pants, jackets, and helmet covers. The ocean is a bright, glittering gray, and there are low clouds over it.

It was a wonderful trip, just the trees gliding by and the rain-wet roads and the hissing of tires, the snapping of chains as we moved up and down our freewheels, up and down, and I realize that I'm absurdly homesick for it. I don't think I had a thought in my head the whole time. All I had was this sense of moving past trees and through rain. All I had was this sense of the bike beneath me, of the spring and tautness of the frame, of the compactness of the load I'd lashed to the racks, all the folds and pockets of my panniers full and tight.

In one picture all you see is a telephoto shot of two overlapping mountains, both steep and forested, and in the center a lighter spot about the size of a penpoint. The spot is a curve in the dirt road twisting up from the valley floor, and in the next shot, if you look closely, you can just make out a tiny figure—the suggestion of a white helmet, a green shirt. That's me. I'm headed up the mountainside, crank by crank, all the way down in my granny gears now, pulling hard on the handlebars. I'm entirely surrounded by green. I'm a tiny speck in the middle distance, smaller than the foregrounded leaves.

I've been to another evening meeting at the Forestry Club Cabin. It's gone later than I thought and, stupidly, I've forgotten to bring a flashlight, so by the time I'm back on the trail headed home it's very dark. It's pretty much what you'd expect. Branches loom out. Things seem bigger or smaller, nearer or farther, than they really are. After a while you begin to notice degrees of darkness, faint and fainter patches of luminosity. Once or twice you stumble and so you begin putting your arms out, feeling for things, and you begin to feel oddly weightless. You can't see your feet or the ground beneath your feet so you have the illusion of floating from the waist up, rustling past the out-stretched leaves.

You think, "So this is what it's like to be in a forest at night," and you realize that you'd be scared if you didn't know your way home. But you do know your way home. After a while you reach a clearing near the edge of the forest. You see lights from houses through the farther trees and you have a sudden clear image of getting into bed and the cool feel of the sheets. You think of the sound the oak tree makes as it moves in the wind outside your bedroom window, the sound of those leaves as you drift off to sleep, and you stop for a moment on the logging road. You stop and turn around and look up and everywhere suddenly there are the stars, the bright sweep of the stars and the brilliant black sky.

The summer before he died Grandfather McRea sat in the tall grass of the blueberry patch while my daughter and I gleaned the last berries. It was one of those perfect July days, clear and still. We were working in the shadow of the hemlock and cedar Grandfather left when he logged the place thirty years ago, but the meadow was full of light, the valley brilliant. Now and then a horse would nicker in the near distance.

I had never known Grandfather to sit down like that. Even in his eighties he could do the work of three men. But weakened then by the bone cancer that killed him he leaned back against a fence post and talked while we worked, clasping his knees like a boy, pausing now and then to look out at the valley, and the light poured down on the meadow, and Maggie and I moved among the cool leaves.

I hadn't expected to take such pleasure in the company of a dying man. The calm I felt in his presence was so deep it had the effect somehow of luxury. There was sadness, too, that final trip, as we prepared for Grandfather's dying, but more than that, unexpectedly, spreading all around us, seeping into us, was a sense of peace and forgetfulness and long, slow releasing. Being with him was like taking the waters at some secret spa.

Later that week on the other side of the state, in Spokane, I had the same sense of silence and release taking Grandma Gottwig to visit the Gottwig graves. The sun was hot on the broad lawns of the cemetery, so I kept Grandma at the fringe, in the shadow of ponderosa, guiding her around sprinkler heads. The air was dry, aromatic. At Aunt Lettie's grave Grandma told the story again: cleaning the house one day, top to bottom—dusting and polishing and folding—and then the fire in the attic that night, flames shooting into the hallway, Lettie dead in the smoke, my six-year-old mother thrown out the upper window, nothing but charred timbers in the morning, blackened furniture. "I don't know," Grandma always says. "A clean house just didn't seem important after that."

But what I recall most vividly is not the telling of the story but the green of the grass and the stillness of the air and the deep pleasure I felt again, for the second time that week. Being with Grandma was as therapeutic as a morning in the garden or a long afternoon in the library. The lawns swept up the hill, the sun shone down, and Grandma and I made our way among the markers, talking of the big fire.

The rest of the vacation I did my usual babbling about the year's struggles and accomplishments. With everyone else I can't stop trying to justify and impress, my words getting away from me, straining and sticky. The pleasure I felt with Grandfather McRea and Grandma Gottwig came from knowing I had nothing to prove. With them my triumphs and failures flattened out in a larger perspective, fading into some simpler, more elemental plot. I remember thinking, with sudden clarity: here I am, picking blueberries with my daughter, and Grandfather is sitting in the grass. Here I am, helping Grandma Gottwig up the hill, and the sun is all around us. Silences were possible, and in those silences a kind of openness, a kind of exhaling.

Sometimes I have this feeling serving as a Eucharistic Minister at the 5:00 Saturday mass. There are five or six of us usually. We process in with the priest, singing, and bow at the altar. After the homily we bring up the wine and hosts, stand behind the priest during the Lamb of God, and after taking communion ourselves, pick up a plate or cup and move to our assigned position (there's a chart in the back room with x's and o's to show where we should be). If I'm giving the bread, I lift each host, say "the body of Christ," and place it either on the palm or the tongue. If I'm giving the wine, I say "the blood of Christ," offer the cup, then wipe the rim with a purificator, twisting the stem counterclockwise with my right hand. Depending on my position around the altar I can serve up to fifty people. When I'm finished I go to the sanctuary to help drink the remaining wine and lock up the leftover hosts, then we process back out after the announcements, two by two, singing the final hymn.

What's occasionally so satisfying about these rituals is their silence and their scriptedness. I speak rarely, usually with others, and the words I speak are not my own. I say what I am supposed to say, stand where I am supposed to stand, in exactly the same way, week after week, inventing none of it, never improvising. There's a sense of participation. I move around in the dark church, the organ playing, children fussing in the back pews, the smell of candles in the dusk, occasionally nodding to another member of the team, quietly purposeful. All that matters is my presence—all that matters is that someone is performing my task. It could be anyone. But it's me. In a sense the ceremony is entirely impersonal, apart from what I've done or failed to do that day. Yet in spite of that, or because of it maybe, I now and then feel a kind of personal fulfillment at the mass that I never feel in the life I invent and direct.

Most of the members of the team are older, in their sixties and seventies, though there is no pulling of rank. We don't know each other outside of mass, yet there's a camaraderie among us now that we've been working together a couple of years. Our greetings are genuine—Ruth in her gray suit, Mildred in her sweat-pants—Ray in his polyester sports jacket, smelling of aftershave, shoes shined, silver hair combed back behind his hearing aid.

It seems a more or less secular feeling, I should say, even indulgent. It's not that I have spiritual insights usually but simply that I come away at ease. The ritual and the community are a balm, physically pleasurable in a way, reassuring.

From the house I can take a long walk through the forest to a stand of old-growth fir. It's about five or six miles away, in one of the upper northern draws. I pack a lunch and take Max and part of the pleasure comes from the rhythm of long walking— the subsiding of other thoughts and the beginning of the slower, deeper thinking, the emergence of the old themes and images. Max trots on ahead, nosing in the brush. The familiar scenes come into view—the stream, the glimpse of the valley, the little grove of maple.

There's been a lot of hyperbole lately about old-growth forests, and in reaction some of the vested interests have questioned the aesthetic categories. What makes very old trees awe-inspiring rather than decadent? All I know is the visceral response I always have moving from the last clearcut into the old-growth itself, the fact of my instinctive feeling as I come into the deep forest. What I feel is not awe but calm. What I feel moving into the old trees is very close to the calm and release and deepening I felt with Grandfather McRea in the blueberries and Grandma Gottwig at the graves, with Ruth and Mildred and Ray standing around the altar.

The huge fir have something of the dwarfing, humbling effect of a starry sky. I feel small among them, clarified by their indifference. Their growth is so slow, their movement so gradual and massive, they seem fixed and stationary, given, and in the midst of that slowness and solidity I am slowed, too, quieted.

I always eat my lunch on a footbridge overlooking a ravine, my feet dangling. In front of me a giant snag points out of maple and oak, the twisted stem of a fir cracked off in a storm or broken under its own enormous weight. Light filters down in the gap. I sit and eat and look at the snag's scarred wood and enjoy even the absence of bird song, the dusky quiet among the trunks. Deer fern and vanilla leaf grow from the bank of the stream beneath me, where Max comes bounding up, muddy and delighted.

There are hemlock etched on Grandfather's gravestone, and there are hemlock growing in the forest behind the cemetery where he is buried, and I think about those trees now, eating my sandwich, dangling my feet over the bridge. Light filters through the gap, the snag twists out of the maple, and I am thinking of Grandfather buried on the hill, slower even than the trees, as slow as the earth.

VI

I have five favorite walks in the forest behind the house, and I take them over and over again, climbing the same hills and looping back through the same trees week after week. I vary hikes according to the mood and the weather, but the key thing is repetition. I've specialized in this forest, in these trails and views.

Standing at the end of the access trail, facing west, I can go in three directions. Left on the 510 Road (south) there's a nice, flat walk through patchy forest to a dead end at the 1990 clearcut, two miles round trip. Right on the 510 Road (north) there's a three-mile round trip through shelterwood and even-aged stands to Cronemiller Lake, a large pond surrounded by old-growth. Straight ahead and up the hill (west) there are three, more difficult, routes: a short, steep walk to Powderhouse Hill; a fifteen-mile walk past the hill following the ridgeline to Lewisburg Saddle and a 40-acre stretch of old-growth; and the six-mile Powderhouse Loop, a varied circuit of roads and trails over the hill, down into the Soap Creek Valley, and back past the lake.

The easy walks I take in the evenings and when there's little time, the hard ones when I'm freer. I go up the hill when I need the sweep of the valley, to the lake when I need the calming effect of water. The lake is best in the morning and in rain, I think, the Powderhouse Loop in late afternoon on a clear day as the light is changing. All the hikes I know by heart, in all their seasons and aspects, the glens and meadows, the prospects of farms, the lusher forest of the western side, the floor all ferns and thimble-berry, rotting logs.

I've walked the forest in spring when the streams were roaring and the winter wrens racketing. I've walked the forest in sum-mer drought when the fir crinkled like Christmas trees and the air was full of the tang of needles. I've walked the forest in the smokey light of fall slashburns, among yellowing maple—climbed the hill in winter over styrofoam snow, my breath form-ing before me, Mt. Jefferson and the Sisters rising above the val-ley's smog. Weekly now for years, sometimes daily, I've walked and walked this same familiar ground, completed countless cir-cuits, learning each leaf and shadow, each angle.

The forest is both rich and close, which is one reason I keep coming back, tracing and retracing my own steps. The terrain is varied, the fir often deep and full. After a few minutes you'd think you were in the Cascades somewhere, or the Coast Range. And yet the access trail is just a hundred yards up our road, no Winnebago required, no forty-pound pack stuffed with freeze-dried stroganoff. All I need is a coat in the winter or a lunch for the long hikes, Max's leash, and I'm gone, home again for dinner and coffee.

Despite my mortgage and my microwave, despite my row of well-heeled neighbors, I'm situated for walking much as Thoreau was. "I can easily walk any number of miles, commencing at my own doorstep, without passing by any house," as Thoreau says in his last, great essay, *Walking*. "Two or three hours walking will take me to as strange a country as I expect to see." Corvallis now is much like Concord in the nineteenth century, a small town still surrounded by green, and living on the edge, like Thoreau, I can roam out into that greenness just as he did, unencumbered, simply stepping off the porch and going.

Freed of logistics, I devote myself to seeing. I specialize in seeing, in learning this particular place, over time and in various light. Specialization brings intimacy. Specialization brings depth of knowledge: of the natural history of the forest, of the human history—of my own history moving through this landscape, a history of images, of moments.

On the climb up the hill are three huge savannah fir, their jagged branches spiking up like tusks, and I often think of how those branches formed: spreading out hundreds of years ago when the trees grew free in oak savannah, breaking off in the crowding of new, younger trees left alone after the first settlers came and the Kalapuya died out, no longer burning the hillsides. I think, too, of the first day my children scrambled up one of the massive trunks, their running shoes dangling from the lower branches, Max barking up at them—the laughter of my children in those ancient trees, and Barb and me sitting on the bench, unscrewing the thermos and the coffee steaming.

Foresters call such fir "hooters" or "wolf trees," too knotty for lumber. The children call them "Grandfather" trees, thinking of their great-grandfather and the tree farm in Aberdeen, thinking of age, feeling an instinctive respect for age. Walking past the trees on my way up the hill, going in my weekly circles, I recall both names, savor both and the connection I feel to them, a connection possible only because of time, because of repetition.

The vista on top of Powderhouse Hill opened up in the Columbus Day storm of 1962, a hundred-year-old stand flattened by hundred-mile-an-hour winds. Over the years Marv Rowley, the forest's first manager, went from one end of the forest to the other salvaging millions of board feet. Five-year-old trees poke up from the thistle and grasses now, planted by forestry students on summer break after the ground was first burned and poisoned, and I often wonder about those trees as I walk through the opening. I think about the faith of some foresters in the old model of clear-cutting and replanting, clear-cutting and replanting, rotation after rotation—I think about recent challenges to that approach, increasing doubts that reforestation in that simple agricultural way finally works. I wonder if the view will ever be blocked by fir.

I have this knowledge, too, these images: Ruth and Jim and I on the summit New Year's Day glimpsing Mt. Rainier, just the barest outline, like a cloud, over two hundred miles away.

Fourth of July and the fireworks shooting off in Monmouth, Albany, Corvallis, all at once, and behind them, in the Cascades, the lightning of a summer storm.

The burnishing of a late October day, and the valley brown and gold and indistinct beneath me, the light deeper somehow, quickening. The call of hawks. The wind.

One morning I met Phil Hayes at the "New Growth/Old Growth Trail" on the Lewisburg Saddle. Phil is a local leader of the Sierra Club, a wildflower expert, and author of *Hiking In and Around Corvallis*, my bible when I first started walking in these hills. He led me down the path talking and pointing, a little like a Boy Scout leader in his khaki pants and shirt, his floppy hat. "There!" he said, "vanilla leaf," breaking off and crushing one of the broad leaves for the faint, clean smell of the oil. Or there's "fragrant bed straw," a tangle of little vines and small candystripe flowers, sticky to the touch, used by pioneers for matting in beds and pillows. Or there's corn lily (stalks tall as corn, standing up from large, parted leaves), inside out flowers (the pistels and stamens on the outside, the petals inverted)— bane berry, bug bane, Siberian miner's lettuce, common groundsels, spring queens, doll's eyes. "I always used to look up when I was in the forest," Phil said, as we crossed into the big trees, "but then, I started looking down. I was just fascinated by the complexity and structure of the flowers." Ninety percent of the species in an old-growth forest are "small and ephemeral," he told me, and so precious, it seems, important. In the last five years Phil has photographed over seven hundred species of wildflowers just in this particular forest, carefully leveling the camera on the steep slopes and among the ferns, entering the dates and measurements into a graphed notebook.

What I admire about Phil is his fidelity to place, his loyalty to detail. I admire his book of hikes, all the quadrants of forest and hill laid out in maps and grids, directions meticulously explained, every vista labeled.

My ambition is to complete a survey of just this kind of detail and complexity, though my interests are more aesthetic, more tonal and selfish. I imagine myself walking at the top of a topographical map—it's in three dimensions—and each time I walk I move down onto another, deeper layer of the map, enter into a deeper level and walk that line, seeing all the things on that line, holding all the other layers in memory. It's a topographical map of perception, of sensory input.

A humbling is required here. The hikes I take are not as spectacular as the hikes in the Cascades, no peaks looming in the foreground, no dramatic ravines. Going in circles is humbling simply because it excludes other good possibilities. But there are too many possibilities, too many good things it's possible to do, and so I need to choose, as we all need to choose, and perhaps through fidelity to my choice, and thoroughness, and faith, I can experience some measure of the whole through the part, at least over time. "To live within limits," Goethe says, "to want one thing, or a very few things, very much and love them dearly, cling to them, survey them from every angle, become one with them— that is what makes the poet, the artist, the human being." One place is maybe as good as another and so we take a chance: this hill, this trail, this stand of fir. Limiting options focuses energy.

In that discipline and exclusion we gather strength as well as manage time.

It's a matter of love, too, of the specialization natural to love and so the channeling of the energy of love, the attentiveness, the regard.

Once a group of us invited Wendell Berry to speak at the university, and he turned us down by way of a little poem. He regrets, he said, his "abuse of courtesy," but he's long since resigned himself to "doing part of my duty / By neglecting the rest." The "beauty / is in the part I do; / The neglect, I guess, is only true." He'll stay put on his Kentucky farm, and we should stay put, too, in our own given places, our own forests and neighborhoods. The issue is the economy, the ecology, of the person. It's avoiding waste by living on a smaller scale, recycling attention and energy into deeper and deeper layers of growth:

> I hate travel by interstate and travel by air.
> I'm almost not going anywhere.
> A bird who cannot fly and sing,
> I'm almost not doing anything
> That can't be done at home. To your health,
> Friend! Try staying home yourself.

Both personal and global health are involved here, not just the saving of gas and jet fuel but the quieting of the self.

I get bored. I grow tired of the forest and need to get away, and it's always good to travel now and then. Going in circles shouldn't be an excuse for complacency or fear of challenge. Not long ago on a trip north we all went sailing on Bellingham Bay, delighted with the wide stretch of water, with the sun and the water and the swelling sail, so different from the forest and the circling, the repetition of our days. The boat was a forty-nine-foot, ocean-going sloop, and early in the day with the wind full in the genoa jib and the sun breaking through the clouds we humped along at nearly nine knots, the children in life jackets out on the bow, laughing and pointing. Later we dropped anchor by Eliza Island and went crabbing, boiling the catch on the galley stove.

But of course, the argument for travel misses the point. The real adventure is inner, not outer; spiritual, not physical. In the evening we return to harbor (completing a circle, in fact) and the next day drive back toward home (another circle). There's always the return, and the hard question is what to do when you arrive back at your beginning.

Thoreau went in circles. He left Concord only a few times in his life, preferring to walk the adjacent forests and fields. What he celebrates in *Walking* are the virtues of staying home and seeing what's close at hand, the wonders outside our own doors. "There is in fact a sort of harmony discoverable between the capabilities of the landscape within a circle of ten miles' radius, or the limits of an afternoon walk, and the threescore years and ten of human life." It all seems tame and safe at first, self-contained—pedestrian.

And yet *Walking* is the occasion for Thoreau's famous statement that "in Wildness is the preservation of the world." *Walking* contains one of his most powerful calls to adventure and the unrestricted life:

If you are ready to leave father and mother, brother and sister, and wife and child and friends, and never see them again,—if you have paid your debts, and made your will, and settled all your affairs, and are a free man, then you are ready for a walk.

The literal walking of this essay is not a figure for safety and domesticity but for "persevering, never-ending enterprise," for "undying adventure." The "faint-hearted" only take "tours," Thoreau chides, seeming to reject the image of the circle after all. They merely "come round again at evening to the old hearth-side from which they set out," unchanged and unchallenged.

There's no contradiction finally. People like to catch Thoreau in personal inconsistencies in order to let themselves off the hook, dismissing him as a hypocritical extremist unable to follow his own advice. They read *Walden* and *Walking* the way fundamentalists read scripture, stupidly, materialistically, missing the real challenge, the harder call, which is to inner transformation. The new country that real walkers seek out is the country of the spirit, of the heart, and for that journey the forests and the fields are only an initial guide, a conducive backdrop. "We would fain take that walk, never yet taken by us through this actual world, which is perfectly symbolical of the path which we love to travel in the interior and ideal world." We walk externally to travel internally. We walk and walk the "familiar fields" so that sometimes we might find ourselves for a moment at least "in another land than is described in their owners' deeds, as it were in some far-away field on the confines of the actual Concord"—or Corvallis, or Seattle, or Spokane—"where her jurisdiction ceases, and the idea which the word Concord suggests ceases to be suggested." We walk and walk, week after week, go in our circles, so that once or twice in our lives perhaps the well-loved landscape will fade, disappear like "mist," to reveal some deeper, realer country.

It's complicated. There's a paradox. On the one hand there's a natural correspondence between the forest and the forest of the mind, the external essential to the health of the internal. "The same soil is good for men and for trees," as Thoreau believed. "To preserve wild animals implies generally the creation of a forest for them to dwell in or resort to. So it is with man." The physical stimulates the intellectual, in other words—is "food" for it. Something in the beauty and order and otherness of the forest orders and calms and feeds the mind. "A man's health requires as many acres of meadow to his prospect as his farm does loads of muck. There are the strong meats on which he feeds. A town is saved, not more by the righteous men in it than by the woods and swamps that surround it."

Which is to say that not just any setting will do for a walk.

There is something in the forest that draws us deeper and deeper—when we would re-create ourselves, we seek the darkest wood. And yet at the same time the larger ambition of walking is finally to escape the physical world and enter into deeper, more profound relationship. The whole purpose of repetition over time is to penetrate deeply enough into a particular landscape that we can dispense with it, at least temporarily.

This is the central point: this is why it's necessary to go in circles. Adventure is too easy. Traveling to exotic places is just a diversion, a distraction from the real project. On Bellingham Bay the scene does all the work, all the stimulating, at least at first and though only apparently. The real work only we can do, undisturbed by strangeness and excessive, unfamiliar beauty.

The unifying claim of *Walking*, as of *Walden*, is that the present is sacred. "Above all, we can't afford not to live in the present. He is blessed over all mortals who loses no moment of the passing life in remembering the past." What most angers Thoreau in *Walden* is the common assumption that true knowledge can be found only in European culture or in the East or in all the faraway places when in fact it's here and now and all around us, culminating in the present moment. Dreaming of eternity is just a dodge. Displacing truth to the stars means we don't have to exert ourselves where we are. There's only one doorway to adventure and that's our own front door, the one leading up the driveway and into the woods. "We are enabled to apprehend at all what is sublime and noble only by the perpetual instilling and drenching of the reality that surrounds us."

The ordinary is metaphor. The ordinary is the vehicle—the only vehicle—for expressing some other tenor, some other idea or state of being we can only intuit and glimpse (Thoreau believed). The lake or the woods may in their substance carry something of the value or essence of this other, higher thing—the levels commingle maybe—and yet the idea is to *read* the lake, *read* the woods. The idea is to get beyond them.

I look out into deep forest—at a road, say, disappearing into the farthest trees—and that impression of depth and inwardness, of receding shadow, of enclosure, arouses in me the deepest feelings of tenderness and sympathy.

And yet at the same time the instinct is to follow the road, out of sight, somewhere else, to penetrate the enclosure. The instinct is to go to the heart.

What follows from this notion of correspondence, of going to the source, is living in a cabin a mile from town and hoeing beans for a year, staying put and living deeply, centering at the one spot close at hand. What follows is hiking the same trails every week for a year, observing the hills and the groves in the light of the different seasons, learning the names of things. The corollary of correspondence is going in circles, because the culmination of God or the presence of whatever we might believe in is located here and only here. New territory simply preoccupies us with externals. We're too gratified by the externals to read them. We're moving too fast. The details blur.

The Benedictines take a vow of stability. Obedience. Humility. And stability: staying put, staying home, within the enclosure. For the life of their vocation they eat and sleep and study and pray within the confines of the monastery, walking the particular fields, regarding the sky from that perspective alone. Because they believe that God is everywhere, they do not need to seek Him elsewhere. He is here, in the place we happen to be, in the family we've been born into—not in some far off, idealized place but here, within this ring of trees, before this expanse of lawns, among these buildings, with all this particular weather.

"The reason why we do not get anywhere," the Desert Fathers said, "is that we do not know our limits, and we are not patient in carrying on the work we have begun." The vow of stability is an act of faith in immanence, in the value of the individual moment. It's a vow because it's hard. It requires the greatest discipline. It is not the casual by-product of some worldly calm and balance but a steady, patient seeking, a laboring against odds. "Go sit in your cell," the Desert Fathers said, "and your cell will teach you everything."

Thoreau derived the word "sauntering" from the medieval term for pilgrims on their way to the Holy Land, those going "a la Sainte Terre." "It requires a direct dispensation from Heaven to become a walker." All walking can be pilgrimage, and especially walking in circles, I think, when finding the way no longer takes all the attention, when it's no longer the external that needs to be mapped and negotiated.

I don't mean that I experience my walks in the forest with anything like Thoreau's intensity or the deep contemplative attention of the Benedictines—or at least not usually—but simply that the goals and the challenges of my walking have as much to do with me as with the forest itself. There are the physical benefits, if nothing else, as in daily running or swimming laps, just the build-up of strength from repeated exercise, as well as the release of endorphins, a kind of walker's high. There's also the build-up of intellectual muscles, at least—the release of mental endorphins, longer thought waves. The physical frees the mental. That's all I can really say, I guess. Walking the same paths week after week has the effect of meditation or liturgy, where the repetition of words and gestures calms and centers the mind, leads it to some still place. The familiar scenes and zones and the rhythm of my walking are like a mantra. My legs loosen, my breathing becomes regular, and gradually I feel my mind flowing into patterns and chains of association, new each time, sometimes leading to silence altogether by the end of the hike, a comfortable wordlessness, not a thought in my head at all.

Going in circles not only frees me to see the landscape, it frees me from the landscape—or it leads to both kinds of awareness at the same time, inner and outer, held in tension.

I climb over the hill and drop down into the Soap Creek Valley. It's winter and the rain is falling. It's autumn, and in the distance Mary's Peak is clear over all the wide slopes of fir. It's morning. It's evening. An olive-sided flycatcher calls in bright sun. Clouds come up the valley and move through the trees.

One day, coming back by the lake, I see a kingfisher stiff-necked on an alder branch.

Sometimes I'm walking, and it's spring, say, the maples leafing out and greening, and I've been walking a long time, and I find myself for a moment in a reverie—the rhythm of my walking and the sweetness of the air have lulled me into a kind of forget-fulness—and for a moment I have the odd sense that just over the hill, just a little deeper in the forest, there's a clearing and the light is shining through. Deeper in the forest (I imagine) is a fountain. Or a circle of stones. For a moment I have the image of a thatched-roof cottage and smoke coming from the chimney and an old man waiting for me at the door, wearing a funny hat, and he knows my name. Just a little farther.

Or sometimes I've been walking a long time, and I've kept going and going, and all there is is the forest, hill after hill of fir and oak, and suddenly, at the top of another rise—where they've clear-cut, maybe, and I have an opening—I stand and see the whole forest extending off in the distance, nothing but trees as far as I can see, and the expanse and sweep of it take my breath away. I seem to see it all at once, take it in whole, and though the moment doesn't last and later I can't account for what I felt, can't explain that sense of gladness, of recognition, I take the moment away and keep it and for a while at least it helps sustain me.

Life and the Essay Compared to a Forest

Nature is a temple where living pillars

Sometimes let out confused words.

Man passes there through forests of symbols

Which observe him with a familiar gaze.

—Baudelaire

People who believe in the New Forestry talk about forests the way essayists talk about essays. They celebrate the same chaos and complexity.

Not long ago a group of us gathered at the base of a two-hundred-foot Douglas fir in the Andrews Experimental Forest while a renowned mycologist demonstrated the stirrup and pulley system he uses to haul himself to the top. We'd fought through thick salmonberry and maple to get there, jumped and slid and stumbled over rotting logs until we'd finally made the clearing, a gap left by some huge windthrow years ago. All around us were the marvelous trunks of the old-growth fir, vaulting, Gothic, and beneath them the feathery second-growth, the snags and stumps and skeletal saplings, the rotting, cluttered floor. Bill Denison was explaining how by swinging out from a boomlike contraption suspended near the top of the fir he can spend hours gathering and measuring *Lobaria oregana*, a nitrogen-fixing lichen essential for old-growth species. *Lobaria oregana* accounts for one-fifth of the "biomass" of the forest, he was saying. Like other lichens it's the result of a "three-way partnership" involving the fungus, the green alga, and the cyanobacterium. "Myrmecia" is the principal photosynthetic partner; "nostoc" fixes atmospheric nitrogen.

And in the midst of his work now and then, I like to think, when he's hanging there by the ropes and pulleys, suspended in the air, Denison looks up from the plastic bags and the clipboards and sees across acres of old-growth canopy. He sees the whole forest at once, it seems, the whole jumbling chaos of its upper surface. Ancient fir cracks and splays at the crowns, each top breaking off into its own array of branches, so that seen from above an old-growth forest must look a little like a furniture warehouse—or like a choppy sea, maybe, the impression almost of movement sometimes, of crests and swells and constant, changing surface.

People who believe in the New Forestry, or "Holistic Forestry," or "Sustainable Forestry," are always using words and phrases like this, admiringly—"irregular," "random," "disordered," "broken," "wild," "messy," "chaotic," "cluttered." "What forestry has traditionally done is make the forest simple," Jerry Franklin argues, one of the New Forestry's most prominent spokesmen. "What we have to learn to do is make the forest diverse." We have to learn diversity in order to keep the forests productive at all, it turns out. Plantation forestry leaves row after row of fir seedlings, neat and clean and orderly, "in billiard-table condition," and there's a certain satisfaction in that. The problem is that forests can't keep growing over time without the nutrients and material of dead and dying trees and all their attendant cycles. Layers and complexities are necessary for productivity. "We have to leave more material behind," Franklin thinks, "leave trees, snags, patches of reproduction, down logs—leaving a lot more heterogeneity in the cutover areas than we've been doing until this point. What we've got to do is leave our Germanic heritage behind us and cherish a little bit of disorder and chaos in our cutover areas."

Essayists talk just the same way. Essayists are always urging us to leave our Germanic heritage behind and cherish a little bit of disorder and chaos in our lives, and they argue that the form of the essay reflects this disorder and chaos, rightly. Their distinction between the "article" and the "essay" (I'm indebted here to the essays on the essay by my friend Carl Klaus) sounds just like Franklin distinguishing a plantation from an old-growth forest. "The scholars distinguish and mark off their ideas more specifically and in detail," Montaigne says. "I, who cannot see beyond what I have learned from experience, without any system, present my ideas in a general way, and tentatively. As in this: I speak my meaning in disjointed parts." An article or treatise has a billiard-table neatness, a row upon row regularity. But Montaigne follows his "natural and ordinary pace, however off the track it is." He lets himself go as he is, "seeking out change indiscriminately."

Joseph Addison says that some of his writing in *The Spectator* has a certain "regularity and method," but that much of it "runs out into the Wildness of those Compositions, which go by the Name of Essays." Samuel Johnson describes the essay as "an irregular, undigested piece, not a regular, orderly performance."

Howells talks about the essay's "essential liberty"; Chesterton, its "leisure and liberty"; Williams, its "infinite fracture"; Kazin, its "open form"; Hoagland, "its extraordinary flexibility"; Lopate, its "wonderfully tolerant form"; Epstein, its "generous boundaries"; Hardwick, its "open spaces." Adorno claims that "the law of the innermost form of the essay is heresy."

And of course there's Thoreau, the essayist who haunts all others since. He acts out his passion for wilderness in the "extravagance" of his prose. He says what he has to say, not what he "ought," refusing all the niceties, exceeding all the conventions for coherence and decorum. What he demands from others he delivers himself: a sincere account of a particular life, the struggles and contradictions unvarnished.

The essay has the structure of an old-growth forest. It's rough and jagged and variously textured, digressive, splintering, apparently unsystematic, a thickness of multiple and indistinct layers, its vertical structures spatially complex and hierarchical, its species—of words and sentences, of ideas—sometimes wildly diverse and always complexly interacting.

The writer of essays [Samuel Johnson says] escapes many embarrassments to which a larger work would have exposed him: he seldom harasses his reason with long trains of consequences, dims his eyes with the perusal of antiquated volumes, or burthen his memory with great accumulations of preparatory knowledge. A careless glance upon a favourite author, or transient survey of the varieties of life, is sufficient to supply the first hint or seminal idea, which enlarged by the gradual accretion of matter stored in the mind, is by the warmth of fancy easily expanded into the flowers, and sometimes then ripened into fruit.

Essays grow from careless seeds, from the varieties of life, slowly accreting, nourished by the matter stored in the mind, and gradually, intuitively, expand into flowers and fruit—or grow up into the canopy, their crowns bristling like imperfect bottle brushes. All the dead and dying ideas, all the ideas abandoned or modified in the course of the writing, are left lying on the page, not revised away, the tentative conclusions rising up from that previous thinking. The levels are there. Gaps open.

There is a coherence here, a deeper coherence, in the forest and the essay, however intricate and hard to see. Wind falls a huge fir and new trees grow up in the opening, replenishing the forest. Rotting logs enrich the soil with carbon and nitrogen the fir can't live without—or the *Lobaria oregana* in the canopy fall to the forest floor, releasing their necessary nitrogen. Some fungi form symbiotic assocations, called mycorrhizae, with the roots of forest trees, which then stimulate the growth of tiny root hairs, which (in turn) help trees absorb phosphorus, nitrogen, and water from the soil. The spores of these fungi are spread in the droppings of the mice, squirrels, chipmunks, and voles that eat truffles, for example, an important mycorrhizal fungus. If the spores aren't dispersed and so don't come into contact with roots, the mycorrhizal relationships can't develop and the root hairs can't absorb the nutrients and the huge trees die—and only the mice and squirrels can do that dispersing. Spotted owls, in turn, eat the truffle-eating rodents, not only further dispersing spores in their own droppings but keeping the rodent population small enough that it won't eat all the truffles. The owls, then, are not just an indicator species in the sense of symbolically representing larger issues; they're central to the issue themselves. "When you try to pick something out by itself," John Muir once said, "you find it is fastened to something else in the universe." Or as Catherine Caufield puts it, paraphrasing the new jargon of ecology, the forest is a "highly efficient and almost closed system."

Or: the forest is like a paragraph, the sentences ebbing and flowing around the main idea, each sentence subordinate to the next, extending and supporting it, the mind moving back and forth toward greater particularity or generalization as it thinks through the idea, explores its implications—or like the paragraphs in an essay, each one advancing a new idea that is at the same time connected to the one preceding, all of them elaborating and developing the essay's larger theme, some relationships implied, others explicit. The subtleties of the forest are like the essay's subtleties. "My ideas follow one another," Montaigne insists, "but sometimes from a distance, and look at each other, but with a sidelong glance"—like *Lobaria oregana*, invisible from the ground but essential to understanding the functioning of the ecosystem. "It is the inattentive reader who loses my subject, not I. Some word about it will always be found off in a corner, which will not fail to be sufficient, though it takes little room." To read the forest requires a great attentiveness. We need to think about truffle spores, study the vole droppings off in some corner, to grasp the intricate order of things.

It's the way life seems sometimes. Mostly the days are messy and fragmented and the pieces at odds, in conflict, but every once in a while, for a moment, we have the intuition of an intricate order, some sense of reconciling ecologies, within and outside of us. Mowing the lawn we think of our fathers and the power of engines and the conservation of power. Turning off the light in the kids' room, we are flooded with hope. Walking through the neighborhood we look up at the stars.

Details make sense. The trivial has meaning. We look across the broken canopy and it seems to us a sea.

The secret of the forest's coherence and complexity is change.

It's change that unifies and explains all the diversity and levels of structure. "The forest is forever in the process of becoming," Keith Ervin says. "Conifers are constantly struggling through the brush layer of a new clearing. Live trees keep being turned into fodder for the fungi that help to grow new trees. Like Sisyphus forever trying to reach the top of the hill, the forest continually struggles towards climax." Understanding the forest means thinking temporally rather than spatially, and thinking in big swaths of time, 200, 300, 400 years. It means understanding "succession" and "disturbance." Eliot Norse defines succession as "a change over time in an ecosystem's species composition, structure, and biogeochemical processes. It occurs when species establish and claim newly available habitats and are, in turn, succeeded by other species until directional change in species composition stops." A fire blackens a hillside, say. In 10 years alder and fir compete for sun; in 50 the alders are dying in the shade of the fir; in 120 a dense fir canopy keeps light from the floor; in 250 the largest fir are strong and thriving while the weaker are dying from insect and storm, returning their energy to the system. In 500 years an old-growth canopy rises over hemlock and cedar, though even this isn't the final stage in the progress. Even old-growth isn't permanent, since over time, 500, 1,000 years, the fir topple and hemlock and red cedar come to dominate.

Succession is only possible when some kind of "disturbance" makes a resource available. The forest can only grow when something goes "wrong": a tree falls, a fire breaks out, there's a flood or an earthquake or even a logging operation and so some opening is created for new species to begin growing.

Change is at the center of the essay, too. Change is the essay's most fundamental insight. "I do not portray being," Montaigne says, "I portray passing." Experience is too shifting and the self too unstable for the writer to make final claims about either. "The world is but a perennial motion," Montaigne says, sounding much like a new ecologist. "All things in it are in constant motion—the earth, the rocks of the Caucausa, the pyramids of

Egypt—both with the common motion and with their own. Stability itself is nothing but a more languid motion." The ecology of the self is equally various and evolving. "We float between different states of mind; we wish nothing freely, nothing absolutely, nothing constantly," which is why the essay insists on dramatizing the act of the mind in the process of thinking through problems, complete with contradictions and changes of heart, none of the progress deleted. The essay is a story of thinking, a record of "passing," spontaneous and immediate, as if in present tense. And a collection of essays is full of contradictions and changing moods. The essayist, E. B. White says, writes "according to his mood or his subject matter," playing the role of the philosopher one day, the scold or enthusiast the next.

The essay reflects life just as the forest is a metaphor for life. Bill McComb, my acquaintance the wildlife biologist, sent me a memo the other day:

> Consider the human resistance to change. We know that disturbance (change) is integral to the functioning of all ecosystems, yet we humans find change, especially abrupt and large-scale change, unpleasant. Why? Do we perceive change as decreasing stability? Does the change that occurs over time remind us of our own mortality (stability may make us feel that time is not passing or at least not passing as quickly)?

The forest is "a forest of symbols," as Baudelaire says in his poem "Correspondences." The trees "look at us with familiar gaze." You keep falling down in public, losing your balance. You walk through the world's largest shopping mall, oddly delighted. The College of Forestry harvests some of the fir in your backyard, and out of that trauma comes a new knowledge of the natural world and a new sense of political responsibility.

Disturbances create new resources and new things grow. The old falls away, the old orders break into pieces, but out of that new diversity, that breaking up, comes a new, a deeper stability.

The secret to the complexity and coherence of the forest is that even the apparently unimportant things are finally important. Everything matters, however small or ugly—the fungus, the owls, the voles. A single tree may have 60 to 70 million needles, a total of forty-three thousand square feet of leaf surface, all of it collecting moisture and chemical nutrients from the air, creating its own microclimate, a world unto itself. Seed taken from fir and pine just two miles away will not grow. The scientists at the Andrews Forest worry about Congress passing some forestry bill mandating one management regime for all patches of ground, for all forests, because they know that each hectare of forest, each part of each hillside, has its own delicate dynamics of soil and water and vegetation. Micromanagement is as important as macro. The scientists just shake their heads when a district ranger in Tennessee calls and asks how many snags she should distribute per acre, as if there's some simple formula, applicable to all. One of the major problems undermining the effectiveness of the Forest Service is its policy of moving people around every four or five years—a policy that makes local knowledge impossible and local history very difficult to retrieve. Each place has its own spirit.

And this is the strategy of the essay, too, and its philosophy: to think small instead of large, to celebrate the small and insist on it. Orwell explains this best when he rails against the abstractions and obfuscations of political rhetoric:

> Defenseless villages are bombarded from the air, the inhabitants driven out into the country, the cattle machine-gunned, the huts set on fire with incendiary bullets: this is called *pacification*. . . . Such phraseology is needed if one wants to name things without calling up mental pictures of them.

Forestry, like all science, is prone to abstractions—to phrases like "treatments" (cutting) or "manipulations" (cutting). The essay won't have them. It focuses in on the concrete details, shows the writer in her own place and time, situated, trying to say what's important, trying to describe what's right outside the window. "It really comes down to what a man wants from a plate of peas," E. B. White says, reflecting on the prepackaging of food and the expanding technology of kitchens, "and to what peas have it in their power to give." The essay always insists on peas, not "the expanding technology of kitchens." It brings things home, exploring how the grand ideas are connected to our day-to-day lives and rejecting all the high-blown abstractions:

> These nuclear springtimes have a pervasive sadness about them [White writes], the virgin earth having been the victim of rape attacks. This is a smiling morning; I am writing where I can look out at our garden piece, which has been newly har-rowed, ready for planting. The rich brown patch of ground used to bring delight to eye and mind at this fresh season of promise. For me the scene has been spoiled by the maggots that work in the mind. Tomorrow we will have rain, and the rain falling on the garden will carry its cargo of debris from old explosions in distant places. Whether the amount of this freight is great or small, whether it is measurable by the farmer or can only be guessed at, one thing is certain: the character of rain has changed, the joy of watching it soak the waiting earth has been diminished, and the whole meaning and worth of gardens has been brought into question.

The only way White could ever manage was at the microlevel, his own level, the level of his own garden and his own rain. And that's a deeply ecological attitude, it seems to me, a foreshadowing of the New Forestry's celebration of lichen. It's not just that the small things matter to us but that they really matter. To overlook them is to lose the forest, lose the earth.

It may even be that the essay is the only form that can honestly and accurately reflect the complexities and dynamics that the New Forestry is trying to understand. It may be that only the structure of the essay can communicate the larger ecologies, natural and personal. It may be that scientists should be writing essays in addition to articles because in writing them they would be forced into a stance of wonder, humility, tentativeness, attention.

The Sustainable Forestry New Paradigm Working Group at Oregon State argues in its statement of principles that "complex systems must be studied as wholes and cannot be fully understood through a study of separate parts. Approaches need to be holistic, generalist, and rely on synthesis." That means "biological and human environments" should be seen as "one sphere of interest," parts of "one complex ecological system," each influencing the other. Science can't do business as usual, given the complexity of the problems it's facing. It must link thought and feeling, empirical analysis and humanistic reflection. It must understand the human elements of this system, the aesthetic and moral. It must see the parts in context, and in multiple contexts, moving up and down a scale of generalization, and always with some sense of humility and irony. It must come up with new ways of communicating knowledge.

That sounds like the essay to me. The essay is personal. The essay expresses feeling. The essay describes what it feels like to be right here, right now, looking out at this beautifully random, beautifully irregular world. It's tentative, modest, not insisting on systems or structures that it knows would falsify the infinite variety. Its concern is always to understand how the experience of the moment is implicated in the larger ecologies, the play of forces.

Swinging out on the boom, looking across the old-growth canopy, it's the essay Bill Denison is acting out, not the article. In that moment of insight, of joy, he is living by essays.

It's morning as I write this, a summer morning, sunny and clear, and I've just come in from looking at Barb's roses. She was up and dressed already but I stood in the wet grass in my bathrobe and Kmart rubber boots, drinking coffee and admiring. Long shadows fell in the yard, and there was a freshness to things, a newness.

The kids' library books are spread all over the livingroom. The dishes need to be rinsed and stacked and the recycling put out in the garage. The curtain's weave makes a pattern on the computer screen, shimmering as the wind moves in the apple tree, and I am sitting here with a second cup of coffee writing down the sentences that have been coming to me: "It's morning as I write this, a summer morning, and I've just come in from the roses. . . ."

Outside now I hear my oldest son playing basketball in the driveway, the ball pinging on the cement, the rim jolting and rattling, and above that, in the fir and maple on the edge of our forest, the sound of the morning birds, the nuthatches and the chickadees.

I like essays because they honor moments like this. I like essays because they give me permission to think and write about moments like this and a language for doing that thinking and writing.

Essays re-create moments like this—moments when nothing happens and everything happens, moments of fullness, of bonus. Reading them I hear my own voice. Reading them I find confidence in the everyday, in the fragments and the glimpses. Reading and writing essays I discover a form for talking to myself and maybe to others about my own given life.

The kind of essay I mean is personal, a sharing of the writer's own thoughts and experiences, here and now, directly, openly. "I want to be seen here in my simple, natural, ordinary fashion, without straining or artifice," Montaigne says in the preface to his *Essays*—Montaigne, the sixteenth-century Frenchman, father of the essay form. "It is myself that I portray. . . . I am myself the matter of my book." E. B. White says bluntly that the essayist is "congenitally self-centered, . . . sustained by the childish belief that everything he thinks about, everything that happens to him, is of general interest."

The kind of essay I mean is concrete, imagistic, narrative. What the "I" usually chooses to share are pieces of experience. There are ideas and abstractions in essays, too, but these are usually reflections on some particular moment in time, some moment when things seemed more vivid or intense, when things seemed clearer. The main work of the essay is to invoke epiphanies, as clearly and directly as possible. E. B. White returns to the summer lake of his boyhood, bringing his son. Joan Didion goes home to Sacramento, wandering through the familiar rooms, emptying out drawers, visiting the family graves. Lewis Thomas visits the Tucson Zoo and is moved by the sight of beavers and otters at play in a big tank.

The moments the essay describes don't have to be earth-shattering and usually aren't. They're more often what Montaigne calls "domestic and private." He writes about smells and his own kidney stones, how he walks, what he eats, the arrangement of books in his study and the sight of the fields all around. Virginia Woolf describes a moth beating against her window-pane. White describes everything from a chimney fire to his dachshund—the news in the local Maine newspaper, the doings of a raccoon in the tree outside his window, his affection for the old woodstove in the kitchen—his chickens, his geese, his peas—all the little details, all the stuff of his own life.

The language of essays like this is direct and conversational, not fancy. Thomas doesn't write "the general consensus is that insects are to be avoided," but "everyone says, stay away from ants." He uses phrases like "all my days," "hankering," "keep a straight face," "while you're at it," the sort of phrases you use when you're talking and not trying to show what an educated person you are. His sentences are often wonderfully short and direct.

Essays avoid inflated and flowery "literary" language, too, the kind Harold Ross of the *New Yorker* called "writerly." Here's White in "Once More to the Lake":

> It's strange how much you can remember about places once you allow your mind to return into the grooves that lead back. You remember one thing and that suddenly reminds you of another thing. I guess I remembered clearest of all the early mornings, when the lake was cool and motionless, remembered how the bedroom smelled of the lumber it was made of and of the wet woods whose scent entered through the screen.

The passage reads like it's not trying to be evocative and lyrical, even though it is evocative and lyrical. It uses unliterary words like "thing," unliterary phrases like "I guess." It dangles prepositions. The mornings are simply "early," the woods "wet."

Montaigne describes the language of the essay this way:

> The speech I love is a simple, natural speech, the same on paper as in the mouth, not so much dainty and well-combed as vehement and brusque, rather difficult than boring, remote from affectation, irregular, disconnected and bold; each bit making a body in itself; not pedantic, not monkish, not lawyer-like but rather soldierly.

It also seems to me that the speakers in these kinds of essays are more open about their feelings than speakers in most contemporary fiction and poetry. On an explicitness spectrum with academic pieces on one extreme (completely explicit, every idea meticulously explained) and fiction and poetry on the other (all implicit, all image, all requiring interpretation of literary detail), the essay is usually dead center. An essay can verge toward the uninterpreted image but it can't go too far. The essayist is like a poet who goes on to explain, to think aloud with us, about the meaning of the images she's first provided.

Or the essayist is like an academic who admits all the reasons he chose a particular subject and what his own biases are. "Articles" are more explicit about arguments but remain silent about the author's motives and feelings. The essay is upfront about motives and feelings, not pretending objectivity.

You can almost say there's a simple formula for an essay, although these elements are woven and textured and blended in intricate patterns within actual pieces: first the re-creation of a moment, then the thinking aloud about the meaning of that moment. It's a natural movement, I think, close to the way the mind spontaneously perceives the world, first experiencing, then reflecting.

The essays I'm talking about always seem spontaneous, however complex the structures. They give the impression the words are just flowing out, without second-guessing, the thinking and the writing simultaneous. "I let my thoughts run on, weak and lowly as they are, as I produced them, without plastering and sewing up the flaws," Montaigne says. He often writes in the present tense, and he often changes his mind as he goes, admitting he was wrong or vague before, trying out a new idea or image, revising right in front of us. White is always doing this, too, giving a movie of his mind in the act of thinking and feeling and caring: "As I sit here this afternoon in this disheveled room, surrounded by the boxes and bales that hold my undisposable treasure, I feel the onset of melancholy" or "I find this morning that what I most vividly and longingly recall is the sight of my grandson and his little sunburnt sister returning to their kitchen door."

Spontaneity tends toward the open-ended and tentative. In the private act of thinking and rethinking an experience, unguarded and untranslated, we're never quite sure of the truth or if there even is one. All the complexities are fresh and uncensored. "If my mind could gain a firm footing," Montaigne says, "I would not make essays, I would make decisions; but it is always in apprenticeship and on trial." Essay means "trial," "attempt." Its purpose is to dramatize the mind in the process of trying to come up with answers—show that whole movement—all the possible conclusions side by side, all the contradictions and fuzziness admitted. "What does it mean?" Didion stops to ask us at one point in an essay, reviewing the stories she's told us, and her answer is that she doesn't know. "It means nothing manageable." Later she says "I have been trying to think . . . in some abstract way but my mind veers inflexibly toward the particular." The essay is always veering toward the particular. It's always trying to represent the movements of the mind whatever they are, tell the story of whatever patterns of thought are developing, chronologically, honestly, without forcing conclusions.

Spontaneity tends toward the brief and fragmentary. Essays are short, intense efforts at understanding, and they don't try to fit into larger systems or establish deeper coherence. Each one is separate, and one may contradict another or be another try at the same experience from a different angle. There's a quality of improvisation about them, of occasionalness. As White describes him, "the essayist arises in the morning and, if he has work to do, selects his garb from an unusually extensive wardrobe: he can pull on any sort of shirt, be any sort of person, according to his mood or his subject matter—philosopher, scold, jester, raconteur, confidant, pundit, devil's advocate, enthusiast."

It's partly this sense of the unpremeditated and even artless that gives essays their aura of being true, not made up, not fiction. However much essayists shape and interpret experience, the impression is that they are shaping and interpreting real experiences, things that actually happened, recalcitrant, messy things, things not easy to fit into plots, not always subject to fiction's shapes and symmetries.

It's largely this permission to struggle that lets essayists bypass all the unsolvable philosophical questions. Literary theory and philosophy are preoccupied now with showing that there is no reality outside of words, that the "I" and the "self" we think we experience isn't really there at all. But because the convention of the essay is to explore and not prove, describe how things feel and not what they irrefutably are, it "brackets" these tough problems, these real problems, holding them at bay so there is at least a usable fiction about rocks and trees and birds and things—without denying the validity of the philosophical challenges. Essays just don't get into all that.

Of course I'm inclined toward all these qualities of the form for reasons of temperament and personal limitation. I'm congenitally self-centered. Like Didion whenever I've tried to think abstractly I've failed. What seems most pressing and urgent to me are my wife's roses and the long shadows on the lawn. What most moves me is the weave and shimmer of the curtains or the sound of my son bouncing a basketball—although I'm never sure what these images mean or if they mean anything important at all.

But I think I'm like most people. I think what's most urgent and pressing for all of us is the ebb and flow of the everyday. I think what most matters to everyone are the bursts of joy and insight, the moments when the ordinary things stand out more vividly, for whatever reason, and this is why the essay has such power. It affirms what matters to us.

There's something deeply invitational about the essay form, something enabling. We're all so tired of high-pitched sales talk and pompous, dogmatic people trying to impress and exclude us that the apparent unstudiedness, the apparent naturalness of the essay is encouraging. Reading essays we know that we won't have to put up with game playing. Writing them we know that we won't have to play games ourselves.

We're glad as readers not to have to struggle through the pedantic and monkish and well-combed. We trust what seems simple and natural and so we're not always resisting the writer's showing off but believing in the words, entering into the experience they describe. As writers we're relieved of the burden to try to sound like "a writer," and that's the key. Writing for posterity is what shuts most people down. You just can't function very well when you're pretending to be a writer or an intellectual. You get all balled up second-guessing yourself, running to a thesaurus, faking drama.

What's enabling about the essay is that the speakers seem common and ordinary and just like us, bumbling even, a little uncertain. They're always confessing sentimental longings, confusions, frustrations, uncertainties. We're not intimidated. Essayists even go out of their way to be self-deprecating, insisting that the essay is no big deal, not to be taken seriously. Montaigne calls essay-writing that "stupid enterprise" and warns us that we'd be "unreasonable" to waste our time on "so frivolous and vain a subject." White compares essay-writing to bird watching, types of essays to flavors of ice cream. He calls essays "childish," "undisciplined," what he "falls back on" when he can't think what else to do. The essayist is a "second-class citizen" in American literature, he insists, "a short distance down the line" from the novelist and the poet. If you want Nobel Prizes, don't write essays.

We should be on guard with these apologies, partly because they're made in precise, powerful language. They're strategies, however sincere. And yet they have their effect. Essayists try to establish a relationship of equality with readers, make us feel that we're on their level. You don't have to take a graduate seminar to read an essay. The reading will be like the "easiest of conversations with an old friend," as Thomas says of Montaigne. The effect if you're a writer is strong, too. You think: Maybe I can do this myself. Maybe there's something I have to say and a way to say it. Maybe I can put something into words right now, without having to be any wiser or more talented.

Reading and writing are always parallel with the essay, mixed up. One leads to the other. The power and wonder of a good novel is to so completely draw us up into its world through the effectiveness of its characterization and plot and setting that we forget we're reading a novel at all. The power of good poetry is to cast a kind of eloquent spell. The words silence us the way liturgy silences us. We're in the presence of mystery. We don't respond to finishing *Moby-Dick* by plunging right away into our own great American novel. We don't get up from *The Wasteland* or *The Duino Elegies* and start jotting down lines.

But just about everyone's response to reading a good essay is to start composing one of her own, if only in her head. As Virginia Woolf puts it, the essay is "the medium which makes it possible for people of ordinary intelligence to communicate their ideas to the world." The gesture of the essay is not to seem like literature. The gesture is not to seem profound. The form of the essay depends on the impression that it's comfortable, natural, doable. An essay makes us think we can write one.

I don't mean that essays are really easy to write or that they really just pour out, although I think first drafts are often written quickly and in the cracks, in the middle of other work and living. Anyone who's ever tried to write an essay knows how hard it is to sound natural, often because the first words that come out on the page are garbled and pompous and complex. Simplicity is an achievement. You have to work to be natural. In the back of Scott Ellege's fine biography of White there are a number of drafts of a paragraph White wrote on the occasion of the moon walk, each full of crossings out and scribbled words. The drafts get smoother and more natural as they go, the first a little stilted, the last seemingly effortless. James Thurber, White's colleague at the *New Yorker*, said the trick was to make something that was "ground out" look "dashed off." For all his protestations of informality and dashing off, Montaigne revised his own essays in five editions over twenty years, sentence by sentence, tinkering, adding, sometimes completely changing meanings.

Montaigne and White were far from ordinary people in their real lives, however much they tried to seem ordinary in writing—more educated and accomplished than the ordinary person, more traveled, braver. All we ever have are the words on the page, and however natural and honest they seem, they might all be fiction for all we know, complete lies. I'm not sure I can explain the difference between a first-person narrative in a novel and the voice of the essayist sharing thoughts and feelings—not sure anyone could tell the difference without knowing in advance. (The essay *seems* real and not made-up, but so do some novels and short stories.)

It's also true that despite their gestures of unliterariness the great essays are great because of their literary power. Great poetry flows from its simplicity. Beneath the apparent randomness and digressiveness complex patterns emerge. Important ideas ring out, profound and useful.

But there are conventions and there are conventions. Conventions have content, one form distinguished from another, and what distinguishes the form of the essay is its claim of naturalness. What's complex about the essay is that it manages to contain all the usual literary sophistication while also maintaining the impression of modesty and ease. It always gets to the complexities, but by way of the simple and natural. Whatever messiness it explores and admits, personal or philosophical, it always begins and ends in the apparent world, in the world we think we experience day to day, the world of all our loves and longings.

In fact the form of the essay is an argument against intellectual pretention and presumptions of complexity—presumptions of any kind. Its humility is an assertion. Its gestures and tones are a natural consequence of a deep skepticism about systematic forms of knowledge, implicit critiques of any claims to truth and privilege. "We no longer know what things are in truth," Montaigne says, "for nothing comes to us except falsified." There is no "existence that is constant, either of our being or of that of objects. And we, and our judgement, and all mortal things go on flowing and rolling unceasingly. Thus nothing certain can be established." For Montaigne this belief comes both from his knowledge of the advances of science and his experience of the violence and stupidity of the people all around him in sixteenth-century France, people killing and imprisoning and oppressing each other in the name of religion or government or other abstractions. It comes from his experience of the obscure, useless, angels-on-a-pin pedantry of Scholasticism; from the strutting of the usual literary poseurs. We all need to feel "the inanity, the vanity and nothingness of man," he thinks; we all need to understand "that the man who is presumptuous of his knowledge does not yet know what knowledge is."

Tentativeness is necessary when there is no fixed, discernable truth. Humility is necessary when you admit your own finiteness. To be autobiographical and fragmentary is to be precise. It is to admit the bias, the situatedness that is always there whether it's admitted or not, the flawed person behind the abstractions, the basketballs and roses and dirty dishes behind every philosophical pronouncement. All we can finally know is what we know ourselves, individually, from our own experience. "What can *I* know?" Montaigne's famous motto, is a demand for honesty, for empiricism. "If the world complains that I speak too much of myself, I complain that it does not even think of itself"—does not even think of the limits of its own knowledge, the fleeting, partial nature of all insight.

Beneath the humility of the essay is humor; beneath its modesty, satire. Both humor and satire depend on the reduction to the physical, on strategies of exposure. "There is no use our mounting on stilts, for on stilts we must still walk on our own legs. And on the loftiest throne in the world we are still sitting only on our own rump." Essays are about legs and rumps—they call our attention to the pants that even the greatest of us put on one leg at a time.

Simply, the essay tries to tell the truth, which is that there is no "Truth," only the succession of days; there is no scholarly authority, only real people in real places and times, struggling to do what is right; there is no reasonable literary affectation, only the hard work of trying to say what's right.

Or this can be put more positively. The form of the essay comes out of a deep faith in the nature of ordinary experience, a love and celebration of it. "I set forth a humble and inglorious life," Montaigne says. "That does not matter. You can tie up all moral philosophy with a common and private life just as well as with a life of richer stuff. Each man bears the entire form of man's estate." It's not just that the "richer stuff" we think we need is in fact an illusion. It's not just that ordinary life is representative, each person an embodiment of the culture. It's that whatever truth and beauty and knowledge we can have is to be found in the common and the private, in the humble. That's where truth is. The inglorious is glorious after all, once we give up our aspiration for pie in the sky, which in its way sounds a good deal like the philosophy of another great essayist four hundred years later. "Men esteem truth remote," Thoreau says, "in the outskirts of the system, beyond the farthest star, before Adam and after the last man. And in eternity there is indeed something true and sublime. But all these times and places and occasions are now and here. God himself culminates in the present moment, and will never be more divine in the lapse of all the ages. And we are enabled to apprehend at all what is sublime and noble only by the perpetual instilling and drenching of the reality that surrounds us."

Thoreau is wilder, more intense than Montaigne, his belief more mystical, but I think the difference is in degree. Roses and shadows and the sound of birds are all the divinity we can ever know (if not believe in). All that is valuable and good culminates in the present moment and our life's work is the instilling of the reality of these moments.

In a sense what the essay is most skeptical about is skepticism itself. It doesn't want to get caught up in the kind of endless intellectual games and deconstructions that obscure the fullness and sweetness of what we touch and taste and feel. "Reason has taught me," Montaigne says, "that to condemn a thing dogmatically as false and impossible is to assume the advantage of knowing the bounds and limits of God's will and of the power of our mother Nature; and that there is no more notable folly in the world than to reduce these things to the measure of our capacity and competence." Essays are not absolutist in their antiabsolutism. They don't even rule out faith in God. They just insist that this faith is a *faith*, a leap, not a sure knowledge justifying dogmatism and arrogance. It's people the essayists doubt—people and their high-sounding schemes and theories—not nature, not God, not the world we actually experience, however dazed and uncertain.

It's evening as I write this, and I've just come in from a walk around the neighborhood. The air was hot and still and the stars shone over the shake roofs and the fir. The Fowlers' Suburban rose up from their immaculate driveway. Yards smelled of cut grass. Through the open windows I could glimpse the bookshelves and lampshades and glowing TV's of the neighborhood, the man getting up from a chair, the woman standing at the sink, and all around us the familiar darkness of the forest, the dark trees climbing the hills to the north and the west.

It's warm and close in the house now, everyone's asleep, and I'm still struggling with the latest draft of this essay, deleting paragraphs and writing new ones: "It's not just that the 'richer stuff' we think we need is in fact an illusion. It's not just that ordinary life is representative, each person an embodiment of the culture. . . ." I'm trying to get the rhythm of the sentences closer to the way I was talking to myself as I walked up and down the streets, thinking aloud about the essay, gesturing. I'm trying to say what I really mean.

I like essays because they call us to say what we mean, to record our truest rhythms. I like essays because they give us all permission to describe the night sky and the summer air and the glimpses through windows. I like essays because they sometimes come close to capturing the way we talk to ourselves on evening walks, beneath the stars, among the rhododendron and the bark dust.

III

I am riding my bike to Independence. It's a hot June day, my birthday, and the hay fields and the corn fields spread out on either side of me. I am a figure moving through a broad, rolling landscape of fields and farms, and the wind is coming from the north. The sky is clear. A red-tail hawk circles. Irrigation sprinklers shoot out streams of dirty water and I can smell the wetness of the corn stalks and of the ground beneath them, and I can hear the water hitting the leaves, and the papery sound of the leaves flapping.

It's a clean, smooth highway. The shoulders are narrow but there's little traffic, only a pickup now and then, and I can see them quite a ways off in the little rearview mirror on my helmet. There's nothing to come for out here, no beach or mountain peaks. The road is just a way of moving through the fields, an infrequent means to ordinary ends, and so being on it now, going slow and looking out at the corn, I have the world all to myself. I am alone on the broad valley floor, and now the corn is high on either side of me, blocking the view, and now there's a wide field of daisies, an acre of daisies, and beyond it a willow brake, the glimpse now of a pond.

I'm averaging, I think, about ten miles an hour. The fields are rolling past me, slowly, and I can see what's in them. I discriminate the rows. I watch the wind bending the hay. I am going slowly enough to be *in* the landscape as much as moving through it, to be a part of it. The hot fields envelope me, and the hot air, and the silence of the hot air.

At the top of a rise I stop to drink from my water bottle and eat a Fig Newton. It's ten more miles to Independence, and I think of the old-fashioned drugstore there and the fountain with the old-fashioned counter and stools, and I think of ice hitting the bottom of a glass and root beer fizzing. But I think of the road, too, the long, clean road, and the yellow fields all around it, I think of the valley, and I'm glad for that, too. The sweat drips from my face and the cows lie down in the pastures and all I am is a figure moving through a landscape. All there is is the sky and the road and the papery leaves of corn, row after row.

I helped start a writing group a year or so ago. Seven of us now meet each week to share what we're writing, a forester, a philosopher, an agricultural economist, a poet, two administrative assistants, and me. I was having coffee with a member of the group over at the Beanery one morning, drinking house blend, and he used a phrase to describe how he'd been thinking. He said, "All week I've been looking for the essay."

Looking for the essay. Once you start reading and writing essays you start looking at your experience differently. You start looking for the important moments to detach themselves, to rise above the other moments, and when they do, as they always do, you recognize them as important, as meaningful, and you celebrate them. You hold them in your mind and reflect on them during the week—whether or not you write them down or whether what you write is good. Looking for essays is a way of thinking, too, and what that way of thinking involves most fundamentally is an appreciation for how life speaks to us through moments. A writing group is a way of sharing those moments, those times when we've been moved or slowed or quieted—a ride to Independence, maybe, and the fields all around, or a camping trip along a river, or thoughts listening to a piece of music some evening. The writing is not important in itself but as a conduit for sharing lives.

"I meditate on any satisfaction," Montaigne says. "I do not skim over it, I sound it, and bend my reason, now grown peevish and hard to please, to welcome it." The essayist sounds experience. The essayist welcomes experience. "When I dance, I dance;

when I sleep, I sleep; and yes, when I walk alone in a beautiful orchard, if my thoughts have been dwelling on extraneous incidents for some part of the time, for some other part I bring them back to the walk, to the orchard, to the sweetness of this solitude, and to me." Thinking like an essayist means meditating on orchards and walks and the sweetness of solitude, meditating on the self, finally, and what it chooses from all the welter of experience to make meaningful—or what experience chooses to disclose to the self, to open up, if only for an instant.

You can live by essays, I think, which is to say that you can live your life as if it is an essay, or a series of essays, internalizing the form as a guide for thinking and acting. Living by essays means going slow enough to see the hawks and feel the contours of the land. "The value of the soul consists not in flying high but in an orderly pace," Montaigne says. You set aside time. You pay attention. It's enough to get there and back and see what you can see, and you can see a lot, finally. Speed doesn't blur what you pass. What matters is the wind bending the hay and the smell of water on the dry fields, the sound of water shooting out of the irrigation sprinklers—the thousand, thousand details, not the easy abstractions.

Living by essays means not necessarily going far but staying close to home and looking deeply at the familiar landscapes for the meaning you overlook when you're going too fast on the way to somewhere more novel and exotic. That's where meaning always is, here and now, in this time and place. "We seek other conditions because we do not know our own," Montaigne says, "and go outside ourselves because we do not know what it is like inside." The beaches and the mountain trails are crowded with view hunters and thrill seekers, and their stereos are playing. Everything smells of suntan lotion. Solitude is close at hand, on the way to Independence—there's no one but you and the occasional hay truck, and there's a smoothness to things, a clarity.

Living by essays means living in modesty and moderation, to

put this another way. It means trying to stay in balance. You try always for equilibrium, not to be in excess of the moment. You ride to Independence, not Sublimity, and you go at your own speed. You go in circles, putting in the miles and tiring out your legs but returning home that evening to birthday cake and party hats and your own cool bed that night.

Living by essays means not taking yourself too seriously. You know that your "conceptions and judgment move only by groping, staggering, stumbling, and blundering" and you don't try to cover that up. The more you look the more "patchwork" and shifting and flawed you find yourself, and you learn to admit that. You try to be honest and direct, saying what you mean instead of what you think you ought to mean. You confess to doubts and silly longings, childish hopes, cliches, contradictions, changes of heart, the simple notions beneath the high-sounding phrases, the leaps of faith.

But living by essays also means trusting yourself. You can't live someone else's life and there's no use putting on airs. Yours are the only legs that can take you to Independence. It's your head that will bow into the wind and your bottom that will chafe. The point is both that the only experience you have is your own— and that this experience is instructive, worthwhile. It reveals all the truths you need. "The life of Caesar has no more to show us than our own; an emperor's or an ordinary man's, it is still a life, subject to all human accidents. Let us only listen: we tell ourselves all we most need."

Living by essays means not taking others too seriously, either. You know from examining your own life and mind that "the man who is presumptuous of his knowledge does not yet know what knowledge is." You know from your own life that "only fools are certain." The world is full of pomposity and arrogance. People don't hesitate to tell you what's true or what you ought to think, but living by essays means resisting the pressure and oppressiveness of such people by continually calling to mind the uncertain grounds of knowledge and the situatedness of all knowers.

Living by essays means being skeptical of all claims to authority and truth, rejecting any idea that you haven't actually experienced yourself, tested in your own life—because ordinary, day-to-day experience is always more complex than abstraction, always more "erratic" and "dangerous" than ideas, and more beautiful. No dogmatism is justified because no one knows the truth. We should have nothing to say "absolutely, simply, solidly."

And yet at the same time living by essays means letting your feelings come into play, not just your thinking. The essay is a method both of skepticism and joy. "We must judge with more reverence the infinite power of nature," and of everything we experience, because it's often reverence that experience invokes, it's often the sense of infinite power, and we need to acknowledge those feelings even when we can't demonstrate them or determine with certainty their source.

Living by essays means recognizing how shifting and unstable and mutable life is—accepting and celebrating that, reveling in it. It means not reaching for grand systems or reconciling truths but accepting the fragmentary, fleeting nature of things, letting go to the chaos and trusting God for the rest. It means accepting the occasional glimpses of order and not striving for more than that, and it means remaining open to those glimpses, ready. "Both the judging and the judged [are] in continual change and motion." Fields and farms and sky keep flowing past us in all their seasons and aspects, all their changing light, and the best we can do is seize some image of them, for a moment, seize and hold and cherish it, and that's enough. The best we can hope for is to be aware of ourselves for just an instant, aware of ourselves as figures in a larger landscape, figures moving through a larger landscape on the way to somewhere else. It's change, the constant movement of things and the blurring boundaries and edges— it's change that makes us attend to the particular, to the present moment, because that's all we can know with any certainty, what we are experiencing right now, right here. Our history "needs to be adapted to the moment," to this moment, to these fields and this sky.

I don't mean that I possess any of these virtues, except the awareness of the fragmentariness of experience and the limits of my own abilities. Arrogance too easily angers and depresses me. I care too much what others think. Ambition continually unmans me. I am always in a hurry, always longing for some overarching order, some reconciling system, even though I know that that order would eventually prove false. But these are the virtues I strive for, the values I most respect. This is the philosophy of living the form the essay naturally reflects and that you naturally begin understanding, at least, when you internalize that form as a way of seeing and acting.

On the way home from Independence I stop at Ridders Pioneer Cemetery. It's a small square of hay field, about the size of a yard, enclosed by chain link fence. Four huge fir trees screen it from the road, but there's a hand-lettered sign when you look closely, and the gate's unlocked. At the center is a waist-high, blackened gravestone and before it six smaller stones lined up like desks in a classroom. The large stone reads "George Ridder 1828–1892" (left) and "Bertha Ridder 1848–1913" and underneath the inscription, "We Will Meet Again." The smaller stones mark the children's graves: "Mary 1874–1923," "Frank 1885–1932," "Gerard 1872–1890," "Lena 1883–1940," "Adele 1879–1916," "Lizzie 1888–1911." At the bottom of each is the inscription "Gone Home."

The shadows are lengthening now and it's cooler in the shade of the four fir—planted, I realize, about 150 years ago, when the little cemetery was first plotted and squared off from the fields. Sweat evaporates in the breeze and my pulse slows. Looking out south and west across the hay fields I have a clear view of

McDonald Forest and the hillside where I live, ten, fifteen miles away. I can make out the uneven edge of the shelterwood on the ridgeline, the bare spot from the clearcut—all that complexity of trees, all those urgent involvements and routines just a range of hills in the distance, a darkening border on the wide, flat fields.

I think of Barb and the kids waiting for me at home, tying off balloons and putting out the cake. I think of my birthday and of growing older. I think that if God speaks to us at all, if there is a God, He speaks in the ebb and flow of the everyday, in the ordinary stories of our lives day to day—and I think that this has the character of that kind of moment, this stopping to look at the graves and looking up to the hills, this momentary perspective and the feel of the breeze, this lengthening of shadows.

I walk around from grave to grave, writing down the inscriptions. I take a drink from my water bottle and eat an apple. It's not sadness I feel, or elation, but a sense of suspension somehow, of openness. The graves have given the day a kind of symmetry, a kind of dignity, and I'm grateful for that, and for the sense of pleasant fatigue I feel, for the fact of all the miles I've gone. I feel soaked through with sun and heat and the fields everywhere, the corn and the hay and the daisies now and then—they've all seeped into me over the hours of riding, worked their way through my skin and my joints, and I'm feeling pliant again, suppler.

Today I've ridden my bike to Independence, and I've seen the hay bending in the wind and I've heard the corn flapping beneath the sprinklers, and I'm standing now in the Ridders Pioneer Cemetery, cooling off and eating an apple, and in the distance is the hill of trees where I live. Sometimes these things are enough. Sometimes things seem to be in balance, in proportion, and it's these times that mark the years and give them meaning. It's these moments in the fields or in the graveyard that we take away and hold in memory and maybe someday weave into the story of our lives.

The American Land and Life Series

Bachelor Bess: The
Homesteading Letters of
Elizabeth Corey, 1909–1919
Edited by Philip L. Gerber

Edge Effects: Notes from an
Oregon Forest
By Chris Anderson

Great Lakes Lumber on the
Great Plains: The Laird,
Norton Lumber Company
in South Dakota
By John N. Vogel

Hard Places: Reading the
Landscape of America's Historic
Mining Districts
By Richard V. Francaviglia

Living in the Depot: The
Two-Story Railroad Station
By H. Roger Grant

Mapping American Culture
*Edited by Wayne Franklin and
Michael C. Steiner*

Mapping the Invisible Landscape:
Folklore, Writing, and the Sense
of Place
By Kent C. Ryden

Pilots' Directions: The
Transcontinental Airway and
Its History
Edited by William M. Leary

A Rural Carpenter's World: The
Craft in a Nineteenth-Century
New York Township
By Wayne Franklin